D1606833

The Fertile Stars

In November 1980, Voyager 1 stunned scientists with this view of Saturn's complex rings.

The Fertile Stars

Brian O'Leary

EVEREST HOUSE *Publishers, New York*

Library of Congress Cataloging in Publication Data

O'Leary, Brian, 1940-
The fertile stars.

1. Extraterrestrial bases. 2. Space colonies.
3. Space stations—Industrial applications.
I. Title.
TL795.7.048 1980 333.79' 099' 2 80-14489
ISBN 0-89696-079-X

Portions of this book have appeared in Omni *and* Next *magazines.*

Published simultaneously in Canada by
Beaverbooks, Don Mills, Ontario
Manufactured in the United States of America
Designed by Joyce Cameron Weston
First Edition RRD681

To Brian and Erin
and their contemporaries,
who have inherited this world
and many others

ACKNOWLEDGMENTS

The inspiration for this book came from my mother, Mary Mabel O'Leary. Since I was eight years old, when it was obvious that I was interested in space, she has gone to extremes to nurture, support, provide new ideas, send clippings, and listen to my outpourings on issues of honesty versus accommodation. Throughout the past thirty-two years, she has known that space, for me, is the place. At the same time, she understands my sometimes maverick participaton in the human quest for the stars.

An example is this book. She painstakingly edited the transcript of a talk I gave that formed the core of this book, and she wrote letters almost daily suggesting titles and twists that are coming to this day. She is a very special lady. Although this book is dedicated to my children, whose world this book is all about, it all started with Mary Mabel O'Leary.

Inspiration has also come from Gerard K. O'Neill, whose friendship and ideas have created the foundation for a large part of this book. I also thank Bill Reiss, Curtis Kelly, Dolores Lefkowitz, and Fred O'Leary for editorial suggestions and technical help that enhanced this book from its first draft.

While numerous other colleagues, friends, and supporters are not mentioned by name in this book, they are true collaborators at the forefront of planning mankind's greatest adventure.

Contents

Earth is a fragile planet, buckling under the pressures of overpopulation, pollution, and the depletion of its energy, food, and minerals. This book tells us how these pressures are likely to be relieved by the abundant resources available in space.

The space solution may be the only solution to these staggering global problems. The major obstacle is political rather than technical.

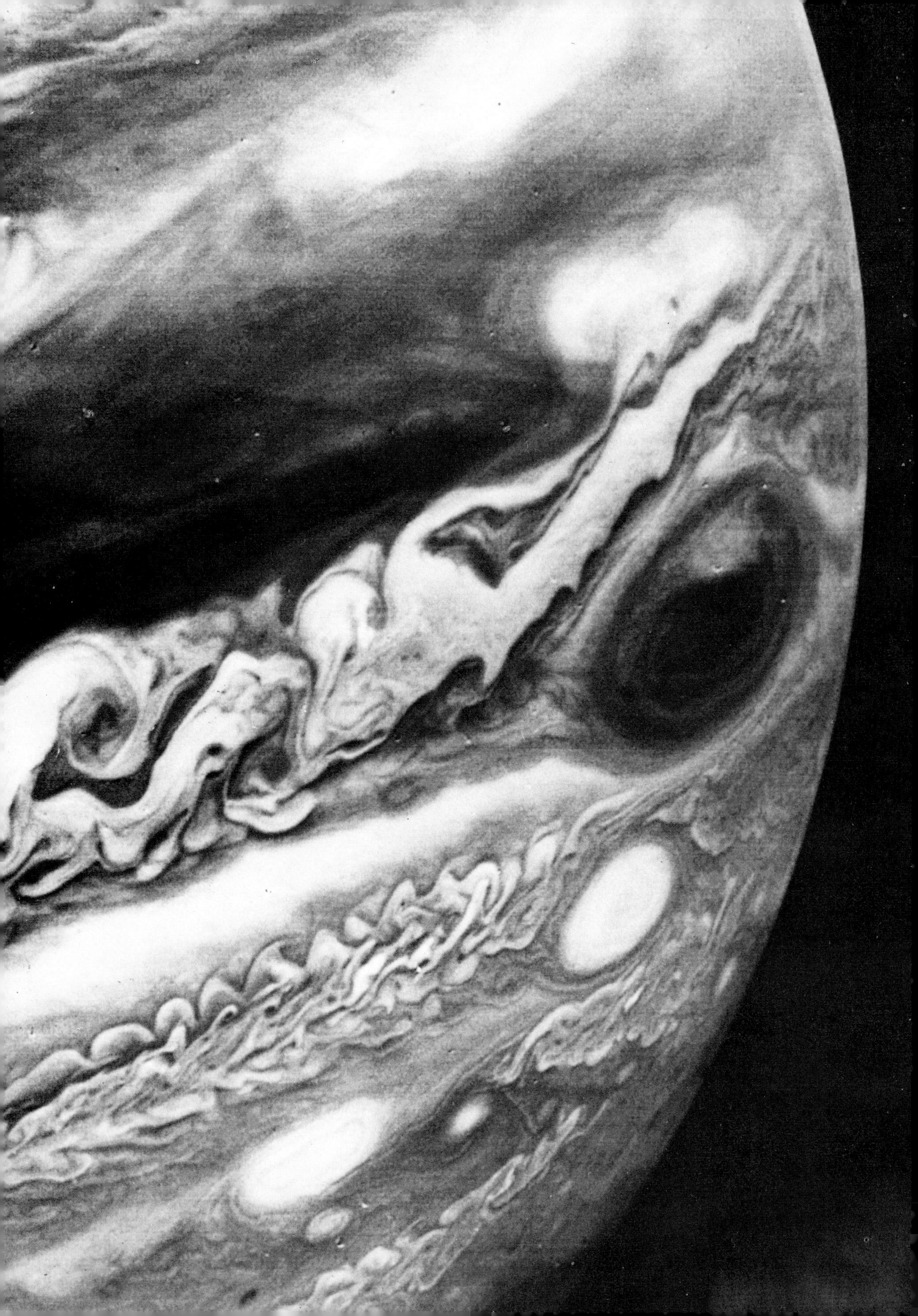

PART I

Realities

This Voyager photograph shows the Great Red Spot of Jupiter (center), a permanent storm bigger than the planet Earth.

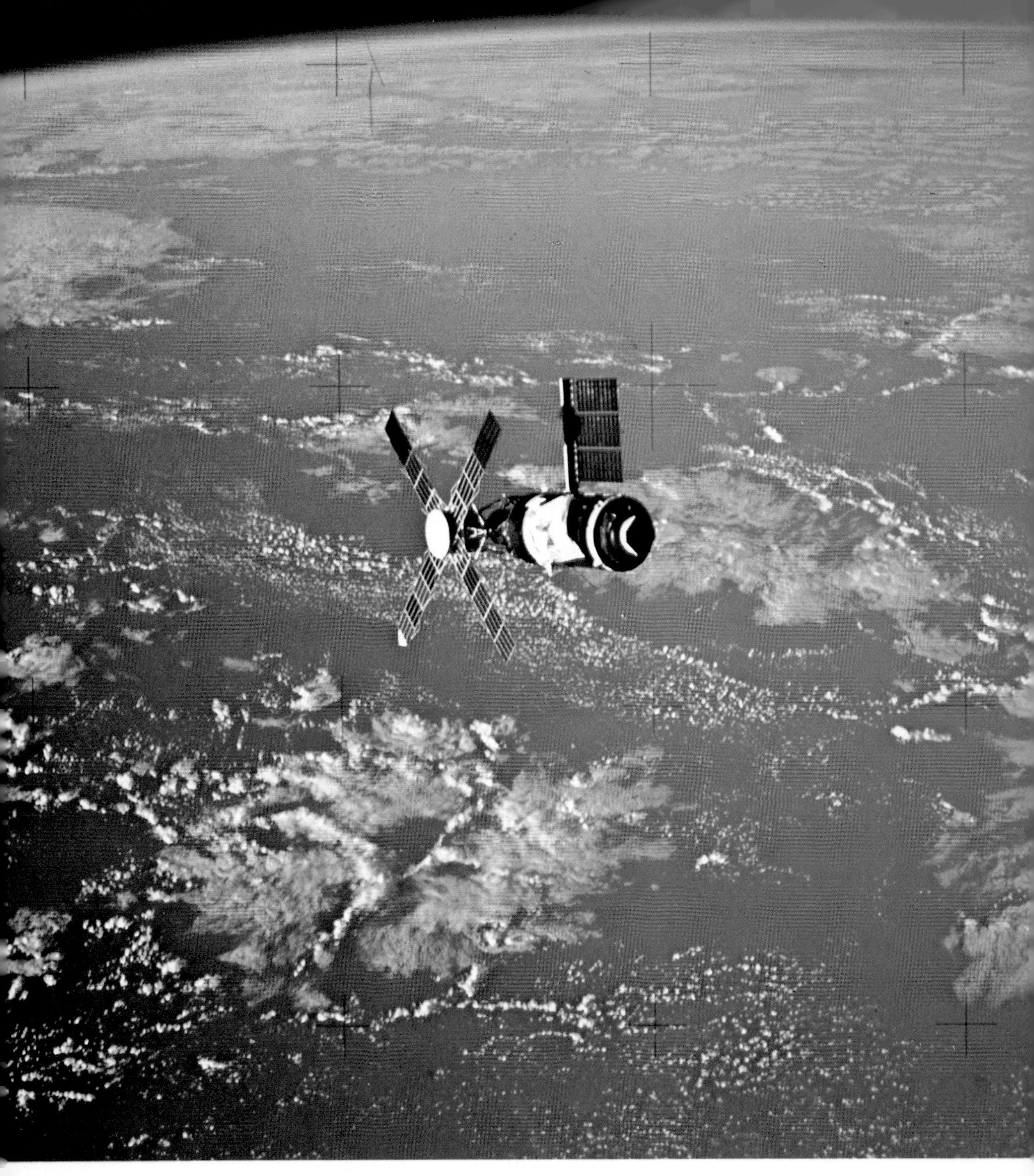

The Skylab space station.

1. 1970-1990: A Cycle of Denial and Myopia

We are on the threshold of the postterrestrial age. On a scale far grander than the pushing of any earthly frontier, we are technologically prepared to thrust ourselves into space. For some of us, the move will be permanent.

The rationale for the bold leap is economic. During the last decade, we have been confronted with increasing scarcities of energy, food, and metals for a growing world population. More than ever, we have polluted and altered our environment. And more than ever, we have seen international tensions develop over the accelerating inequality of resources and balances of trade. The most obvious and ominous example is the importation by the United States of foreign oil at a staggering cash outflow of $50 billion each year.

Over the past decade of increasing scarcity, we have learned that the relationship between economics, energy, food, materials, and the environment can create intertwined global problems that cannot be solved in isolated sectors. And we have learned that there is a clear-cut and feasible answer to all these problems: we can use the resources of the asteroids and energy from the sun, available to us in great abundance in space, to relieve the limits to growth on earth.

The inevitability of the asteroid solution is obvious to an increasing number of scientists and engineers, but it staggers the imaginations of most people. The idea of going into space to mine asteroids creates symptoms of future shock, Alvin Toffler's term for the dizzying disorientation people experience from the premature arrival of the future.

How we choose to respond will be up to each of us. Unfortunately anxiety, confusion, withdrawal, and depression are likely to be the most common symptoms. Extensive psychological testing tells us that one of the first behavioral responses to the stresses of change is denial. "The disturbing fact," said Toffler, "is that the vast majority of people, *including educated and otherwise sophisticated people* (italics mine), find the idea of change so threatening they attempt to deny its existence. Even many people who understand intellectually that change is accelerating have not internalized that knowledge, do not take the critical social fact into account in planning their own personal lives." Toffler documents well the denial among well-informed people of the potential realities of the first telephones, airplanes, automobiles, and other technological breakthroughs.

A major thesis of this book is that the ultimate future shock most of us will be facing in our lives is space shock: we are in space now and there is no moving back; we will soon be expanding enterprises into space at a staggering pace; and all this will profoundly affect our lives. We can choose either to deny this reality and withdraw into old, unadaptive behavior patterns or to learn how we can participate in the great, dizzying new adventure that is beginning to surround us. Denial can only lead us to immobilizing worry and fear.

It is fruitless to make value judgments about whether we *should* go into space, whether mining the moon and asteroids is the *right* thing to do, or whether we *ought* to build space shuttles or receive solar power from space—for we are in space to stay. Given that the economic case for using the resources and energy of space is already sufficiently compelling, as I shall demonstrate, we will probe deeper into space. Maybe not right now, or next year, or even in 1990, but the space solution is inevitable, and, as shocking as it might at first appear, it is too late to debate the right-or-wrong of it.

I also believe that the most likely path to our pervasive and permanent presence off the earth will be precipitated by some event, or series of events, that will likely take place during the 1980s. It will be a major international crisis or a bottoming out of our current social malaise. I will present the case that going into space appears to be the *only* logical remedy of sufficient scale to handle the most fundamental problems facing humanity.

Toffler also refers to "political myopia" in *Future Shock.* Economists, he wrote, "see technological advance as a simple, nonrevolutionary extension of the known. Born of scarcity, trained to think in terms of limited resources," economists equate material needs with human needs and create the basis for the day-to-day, incremental decisions of our current age. But a blend of profound social change and virtually unlimited resources in space, as we shall see, invalidates most of the laws of contemporary economics.

Because conventional, outmoded economics guide our current public policies, because of the short terms of elective offices, and because of the acceleration of events, democracies more than ever before seem to be unable to deal effectively with the most basic long-term problems of mankind. Shortsighted planning is the norm in the nation best equipped to develop new ideas: the United States.

Leaders and bureaucrats repeatedly assume that because a given idea is newer than another, or has just reached a level of consciousness, the results of this idea must be further off in time. One is told to come back in the twenty-first century. Taking these linear projections into the future at face value can get discouraging. In these times particularly, bureaucratic red tape too often snarls the development of sound, creative concepts, reinforcing a sense of denial. Yet the reality persists and dwells in each of us, though sometimes latently.

The decade of the 1970s has been a period of insecurity and self-doubt in the United States, of a decline in economic and moral will, and the feeling seems to be spreading throughout the world. As we enter the last two decades of the century, we see that problems demanding lasting solutions, instead of being solved, lead to relentless discussion, debate, dissection, and soul-searching. Often with the best intentions, our leaders come up with short-term, incremental Band-Aids in order to buy some time until the next crisis hits—or until the next election. Agencies are established, committees formed, reports filed, regulations revised, legislation passed, speeches made, and editorials written. The results are usually complex and inconclusive and treat the symptoms rather than the causes.

More lasting solutions are buried in a political tower of babel. It may take years, particularly during pessimistic times, for such long-term concepts to surface in the public consciousness.

The world energy problem has staggering proportions and cries out for a solution. The American people and their leaders may be just beginning to acknowledge that the roots of the problem lie deeper than occasional shortages, price increases, regulation, and other vicissitudes of political and economic systems. Unlike the hostages in Iran, we are 220 million energy hostages who will not be released during the next several years.

In his energy message to Congress three months after his inauguration, ex-President Carter called the attack on the energy problem "the moral equivalent of war." He said that the nation's oil supply was running dangerously low and that grave international instabilities would result from our increasing dependence on foreign oil, which itself will run out in time. Carter proposed incentives for energy conservation, but he stopped short of proposing the development of alternatives. Bending to special interests, Congress has all but decimated the plan, whatever its worth. Meanwhile, the government has moved on to other things, and business-as-usual and crisis management prevail. Denial is rampant. Apart from occasional flurries of rhetoric, American leaders provide no sense of vision, no inspiration.

Our hunger for energy and our dependence on foreign oil are grim facts of life, but there is little public evidence that we are any closer to confronting them. During a breakfast that I attended in Providence, Rhode Island, in 1979, President Carter lamented these problems and blamed them on a recalcitrant Congress, bending to special interests, but he was unable to move or inspire his audience.

"We are energy alcoholics," said *The New York Times* in support of Carter's postinaugural plan to raise energy prices in order to conserve energy. *The New Yorker* stated that the President "was acting in response to a great and obvious global peril, which stems from the indisputable fact that the Earth's natural resources are finite and are being consumed at an ever-increasing rate, and which has placed the independence of American foreign policy in doubt and has even disrupted the routine of daily life by forcing the evacuation of homes for lack of heating, and creating, among other instances of serious hardship, the layoff of millions of workers in factories for which fuel ran out."

At the same time, *The New Republic* called Carter's undiluted energy plan "timid" because of its "failure to call forth the creativity and intelligence of Americans to develop alternative energy sources. . . . We must act with zest and dedication to fund the scientific muse in its efforts to find a way out of the deepening crisis. . . . The Western experience rests on the marriage of science and technology; its ethos is not contemplative, but expansive. *Our future must even allow for extravagance and waste as the Italian Renaissance did, and the English 19th century* (italics mine). If we are without aims beyond possibility, we are bereft of aims that inspire. Right now, it seems, bureaucrats and technocrats are busy rebinding Prometheus to his fetters and taking his fire away besides. It could not be otherwise. For civilization is the imagination externalized, and the contemporary imagination, more than ever shaped by public officials and their journalistic pulpits, is punitive and impoverished."

The tradition of short-sighted planning appears to be continuing under the new Reagan administration, from all early indications.

The Malthusian dilemma of a finite earth has been a common theme in scholarly circles for some time. It is now upon all of us. Two decades ago, Aldous Huxley wrote, "The problem of rapidly increasing numbers in relation to natural resources, to social stability and to the well-being of individuals—that is the central problem of mankind; and it will remain the central problem certainly for another century. . . ."

Where are we headed on this finite planet? How are we going to handle shortages during our lifetime? What are we bequeathing to our children?

There can be little disagreement that the planet Earth is in a great deal of trouble, that it is on the verge of a nervous breakdown, as Daniel Patrick Moynihan and Mort Sahl have put it. University of Wisconsin sociologist L. Suhm wrote, "We are going through a period as traumatic as the evolution of man's predecessors from sea creatures to land creatures. . . . Those who can adapt will; those who can't will go on surviving somehow at a lower level of development or will perish—washed up on the shores."

Each individual has a choice to make: ride the new wave of culture that will breathlessly catapult us into space or be washed complacently ashore. Mankind has enormous problems, and solutions short of the space solution are probably not adequate. "Small-is-beautiful," energy conservation, and changes in political and social structures are embraced as possible solu-

Sunsats extending several miles across may solve the world's energy shortage.

The inside of this space habitat, as seen reflected from a huge mirror, houses 10,000 people.

tions, but they are simply inadequate or cannot be implemented realistically. The result of present-day thinking can only be the fulfillment of the nightmares of doomsday prophets, who are writing and selling books at an unprecedented rate.

We can choose to wallow in grief or be immobilized by fears: "Self-indulgent despair," wrote Toffler, "is a highly salable literary commodity today. Yet despair is not merely a refuge for irresponsibility; it is mystified," referring to the way that we have virtually made a religion of despair. Alternatively, we can choose to keep pace with galloping events and share in both the blessings and anxieties of space shock.

Nobody has yet been able to demonstrate convincingly how we can turn around these fears, how we can transcend the decline and sickness that many of us feel. This book attempts to provide an answer, but we can share in it only if we are prepared to embrace it emotionally as well as intellectually. The feelings are there, in latent form, among at least a million people, mostly young and forward-looking, who read the magazines *Omni*, *Science 80*, *Next*, *Quest 81*, *Science Digest*, and *Discover*. They ignore the denials and myopia of contemporary communications; they choose expanding future-oriented careers; and they look with wonder at the extraordinary beauty of the Voyager pictures of Jupiter's Galilean satellites. These people are prepared to jump in as the pioneers of the postterrestrial age.

The ultimate solution to the energy crisis is to seek our natural resources in space. It is remarkable and fortuitous that the space solution can be accomplished with existing technology over the next ten to twenty years, step-by-step, without irrevocable commitments of billions of dollars. The cost would be many times less than that earmarked for the alternative short-term solution: the use of more coal and uranium. Satellite solar-power plants built from asteroids or the moon may be able to supply the earth with abundant electricity in economic competition with existing sources, probably at a far lower environmental cost. Food and raw materials can be obtained from space at lower cost and expand the terrestrial supply enormously.

Natural-gas shortages, high oil prices, Middle Eastern political and economic blackmail, gas-station lines, coal strikes, nuclear proliferation, famine, starvation, the depletion of fossil fuels and other non-renewable resources, the rape of the environment—these are problems that demand lasting solutions. Necessity being the mother of invention, the space solution *will* become reality.

The eventual outcome of unlocking the resources of space is the spread of humanity throughout the solar system and stars. History will undoubtedly judge the latter event to be of more lasting significance than the original intentions. The movement into space may ultimately become the key to human survival, for it begins at a time when nearly all of the planet's resources are being depleted.

Because I propose that the space solution is not only inevitable but imminent, this book will probably create some controversy. If skeptical readers will acknowledge that these events *may* some day take place, that will be a significant step, for it will enable you to gain a perspective beyond conventional wisdom and soften the blow of space shock. Visualizing something you believe *may* happen, even if not for decades, allows you to explore the consequences of a possibility that may not have occurred to you before. You will then be free to explore uncharted territory without considering whether it will happen or whether it is right or wrong.

The main purpose of this book is to reveal that very special perspective, to share the excitement that awaits all of us in space, and to guide you through the troubled time of transition that we must experience before we get there.

"The earth is the cradle of man," wrote the Soviet schoolmaster-visionary Konstantin Tsiolkovsky eighty years ago, "but man cannot stay in the cradle forever." The anxiety we now feel, the convulsive pains of growing up, will probably get worse before they get better. But what lies beyond is so immeasurably stimulating, rich, and vast that the pain will have been worth it and soon forgotten.

Mankind is enmeshed in a cycle of denial and myopia that is intensifying. The reality is that we are in a deep decline that will worsen and then climax with a shocking event, probably of a magnitude greater than Pearl Harbor, which helped spawn the atomic bomb project, or the Cold War, Sputnik, or the Bay of Pigs invasion, which sent Americans to the surface of the moon in eight short years after John F. Kennedy set the goal.

This book departs from the writings of the doomsday prophets in one major respect: at the other side of high anxiety, I see relief and excitement and good times, not obliteration, despair, or another Stone Age. In my opinion, the gathering storm of crisis is the trigger that will propel us into the space solution. But the space solution will carry with it space shock, because the pace of events will be far faster than many of us will be able to handle.

"In our society at present," wrote Harvard psychiatrist Erik Erikson, "the 'natural course of events' is precisely that the rate of change should continue to accelerate up to the as-yet-unreached limits of human and institutional adaptability." What are those limits? In the coming years they will be tested as never before. Space shock, the ultimate shock has already arrived and will accelerate at a dizzying pace. The good news is that there are ways of coping with events through increased awareness of their meaning, direct participation by more people, the vicarious pleasure of exploring new frontiers, and the knowledge that economic recovery is on its way at long last.

This book explores in detail the genesis of space shock, examines the exciting events projected in the future, guides us around the denials and myopia that pervade the times, describes the likely crises of the 1980s, and concludes with how mankind's permanent presence in space will profoundly affect every one of us. Beyond the current short-lived cycle of denial and myopia lies the postterrestrial age. It has already begun.

2. *The Next Ten Years in Space*

More than any year during its 23-year history, 1981 is a pivotal year for the National Aeronautics and Space Administration (NASA). The entire United States civilian space program, and other projects as well, hang on the successful launching and orbital testing of the space shuttle Columbia, the first rocket plane ever to go into space.

What lies on the other side of the first launch, which NASA plans for the spring, is either a radically revitalized space program or a sick space program. One way or the other, the space shuttle promises eventually to be the key to the American space effort during the 1980s.

(Left) The space shuttle jettisons its two boosters as it heads toward space. (Below) Astronauts aboard the shuttle will place large cargoes, like this telescope, into orbit.

A typical shuttle mission will begin with a loud, fiery launch from either Cape Canaveral, Florida (for NASA, commercial, foreign, and some defense satellites), or Vandenberg Air Force Base in California (for defense satellites needing highly inclined orbits around the earth). The DC-9-sized, squat delta-wing rocket will sit piggyback on a larger external fuel tank, which will feed liquid oxygen and liquid hydrogen into the space shuttle's main engines. Two solid-fueled tanks strapped onto the liquid tank will help give the shuttle an initial boost before they are jettisoned and splash into the ocean 160 miles downrange, where they can be towed ashore by barge and be used again.

As the shuttle approaches orbit, it will jettison its external tank, which will burn up on reentry into the earth's atmosphere over the Indian Ocean. The typical orbit of the shuttle will be 160 nautical miles high, and the spaceship will stay in space from a few hours to 30 days. Its main purpose will be to deliver and retrieve satellites that can fit into its cylindrical cargo bay, 15 feet in diameter and 60 feet long. An impressive 25 tons can be packed into the cargo bay.

The shuttle's mission ends when it descends through the atmosphere and hurtles at 215 knots along one of two special three-mile-long landing strips at the NASA Kennedy Space Center or at Vandenberg. The shuttle will then go into a hangar for several weeks' refurbishment before its next flight into space. The new spaceship will carry up to seven people, two of them pilots who will maneuver the vehicle and override any errant automatic systems.

In recent years, fiscal austerity has kept NASA's plans at bay, so much so that the agency has felt compelled to keep a low profile, put all its eggs in the shuttle basket, and cut back severely on other projects. Over the past decade, tight-fisted federal administrations and the Office of Management and Budget have so severely limited NASA's spending that it has had to go back to Congress twice for more money to keep the shuttle program alive. As a result, NASA has delayed the shuttle's first orbital flight several times (from 1978 to 1981) and has adopted the restrictive philosophy of "success-oriented management," wherein everything is assumed to go all right.

It hasn't. Thermal protection tiles have fallen off, engines have exploded in test-firings, and parts have been found to be missing. The results have been nearly devastating: continual delays, cost escalations, narrowing safety margins, embarrassment. The first 1981 flight is certain to be a cliffhanger for Columbia astronauts John Young and Robert Crippen, for NASA, and for all the rest of us.*

Newspaper headlines about the shuttle's troubles, however, tend to distract us from looking at what awaits us in space during the 1980s.

**This was written before the highly successful* Columbia *launch which insures us a clear pathway to the fertile stars.*

NASA has spent the vast bulk of its $9 billion in development money for the shuttle, and it is unlikely that the program will be scrapped entirely, even if serious problems crop up later this year.

Moreover, military and Soviet pressues are likely to sustain the shuttle program during its first years of shakedown. For better or worse, all the evidence points to a reinvigorated space race, with the shuttle serving American needs. Ultimately the shuttle will be able to catapult humanity toward opportunities that stagger the imagination: energy, food, metals, and industrial products from space, and the chance for ordinary people to colonize space and explore the universe.

If we assume, then, that it is likely that the shuttle will survive hard times, what can we say about what will happen in space over the next ten years? The answer is that we can be nearly 100 percent certain about what the United States and European countries will be doing between now and 1985; we have some information about what the Russians are planning over the next few years; and we can create a range of scenarios about where these activities might lead over the second half of the decade.

Because it takes five years to prepare a spacecraft, NASA's shuttle missions between now and 1985 are "cast in concrete." If the shuttle's test flights take place satisfactorily and on schedule, we will see during the three years following September 1982 40 launches, first involving two, then three, and finally four shuttle orbiters. The schedule will slip if the test flights slip, a very definite possibility, but the same missions will still fly.

The 40-mission plan is impressive. The features that make the shuttle unique are its enormous cargo capacity, its reusability, and its ability to visit satellites orbiting up to 600 miles high for repairs or the replacement of modules. It can even pluck an entire satellite from space and return it to earth. Using the second shuttle orbiter, the Challenger, NASA plans to do just that for the Solar Maximum Mission satellite in December 1982. Launched February 1980, this spacecraft's telescopes are currently monitoring the spectacular eruptions coming from the sun during its sunspot peak.

A description of the 40 NASA operational shuttle launches between 1982 and 1985 appears in the document "STS (Space Transportation System) Flight Assignment Baseline," issued in June 1980 by the NASA Johnson Space Center in Houston. The report is a succinct guide to the coming space program, depicting a Heinz-57 variety of payloads being built by commercial groups, NASA, the Pentagon, other government agencies, and foreign governments.

Embellished with NASA abbreviations and acronyms, the document shows cross-sectional sketches of shuttles for each of the 40 missions, with a color-coded drawing of the appropriate satellite(s) placed in the cargo bay. The only exceptions are the payloads of the Air Force, who will be sending ten secret cargos into orbit by shuttle. (They had planned to send 20 cargos, but shuttle troubles forced the Air Force to use its already-tested Titan III for the other ten.)

As we approach 1985, the Department of Defense will become NASA's principal shuttle customer. Large surveillance satellites, such as the Air Force's Big Bird, will dominate the traffic over the next few years. This activity may increase even more if the Senate ratifies the SALT II arms limitation treaty. Enforcement of the SALT treaties is possible only under the watchful eyes of spy satellites, which can detect objects on the earth as small as one meter across!

Much has been said about the earthly benefits that can accrue from images taken from space. The programs already under way concern weather forecasting, hurricane tracking, mapping and charting, land use, agriculture, forestry, pollution monitoring, water resources, mineral resources, petroleum exploration, earthquake fault monitoring, and ocean current patterns. It is difficult to place a value on the benefits of these programs, but it is certainly very high. Although NASA often extols the merits of the shuttle for these earth-resources programs, it is curious that none of the 40 committed shuttle flights will carry any satellite like the Landsat. The situation may change in the late 1980s, when the Department of Commerce hopes to launch advanced Landsats. Meanwhile, NASA will be sending up two more Landsats in 1982 and 1983 with conventional launch vehicles.

The communications industry relies increasingly on satellites for international telephone, radio, and television links, and it has reaped impressive profits as electronic equipment becomes more sophisticated and launch costs go down. Several communications companies have bought shuttle space, and in the early 1980s, a series of new satellites owned by AT&T, Intelsat, RCA, and Hughes will be launched. The governments of India, Indonesia, Canada, and Saudi Arabia have also signed up for communications satellites to be flown aboard the shuttle between 1982 and 1985.

The space comunications industry will undoubtedly continue to grow rapidly. By the early 1990s, according to studies prepared for the U.S. Telephone and Telegraph Corporation, long-distance telecommunications services will have saturated the nation's existing domestic satellite capacity in existing bandwidths. More versatile satellites will be needed to meet the growing demand, and such new technologies as video-conferencing and two-way radio communications using Dick Tracy wrist devices may be commonplace, replacing business travel and normal telephone links.

(Above) Its mission completed, the shuttle then absorbs the searing heat of atmospheric reentry just minutes before landing. (Below) The test space shuttle Enterprise approaches its historic landing after being carried aloft by a Boeing 747.

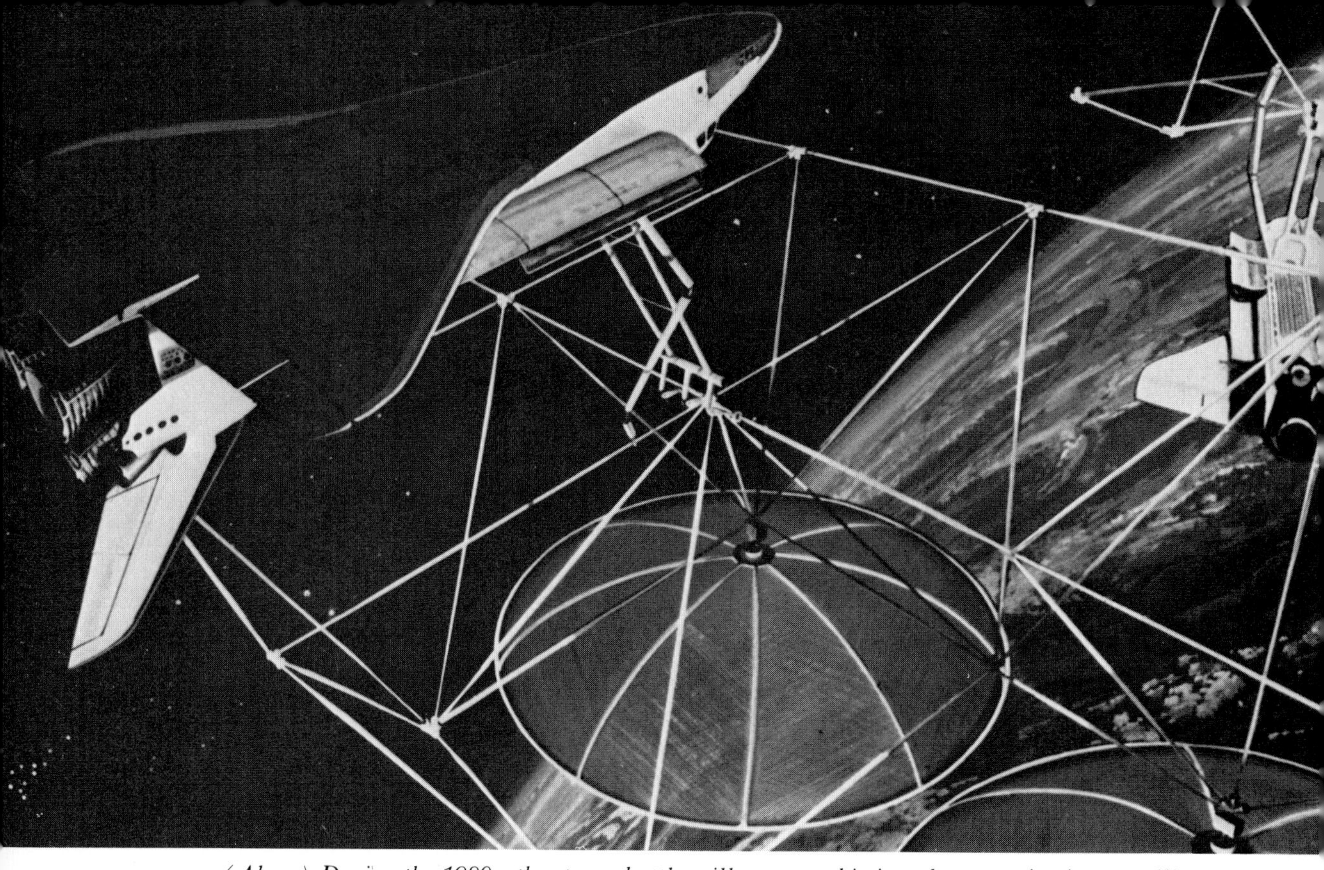

(Above) During the 1980s, the space shuttle will carry sophisticated communications satellites like these into orbit.
(Below) During the mid-1980s, the Space Telescope will revolutionize astronomy.

Europeans will figure prominently in the space-shuttle program. The European Space Agency, a consortium of eleven Western European countries, will be orbiting a manned laboratory called Spacelab. European and American scientists and engineers will ride aloft in Spacelab's shirt-sleeve environment to conduct a battery of medical, scientific, and technical experiments reminiscent of Skylab and Soyuz. The first Spacelab is scheduled for a May 1983 launch, with two West German Spacelabs planned for 1984 and 1985, as well as six American Spacelabs during the same period.

All indicators point to a European renaissance in space, the most recent evidence being their willingness to fly a Halley's Comet flyby mission in 1985, using their own launch vehicle, regardless of what NASA decides to do. (NASA would like to launch a Halley probe on the shuttle, but fiscal pressures have held them back from any commitment.) The Europeans also plan to use the shuttle to place a probe into polar orbit around the sun. Along with a similar American spacecraft, the mission should provide new knowledge about the sun's little-known polar regions, cosmic rays, and magnetic fields.

There is no lack of popular interest in space among the Europeans. In the summer of 1980, I joined other astronauts and astronomers to address Rome's Second International Festival of the Poets, where an audience of several thousand responded with great emotion and enthusiasm. Then I traveled to France, where I discovered that magazines displaying space-related covers were rivaled only by pornography. The August 4, 1980, issue of the popular news magazine *Le Point* featured a six-page article on the practical uses of space. A second prominent magazine, *Espace et Civilisation*, was rich with articles on spaceflight and astrophysics. Conspicuously absent from both magazines, however, was any mention of the United States program, in sharp contrast to the heyday of Apollo. The European attitude seems to be that they can go it alone in space, apart from the occasional purchase of space on the shuttle.

Probably the most exciting shuttle payload will be NASA's Space Telescope, scheduled for an early 1984 launch. Peering to the edge of the universe, the spacecraft's 2.4-meter (95-inch) mirror will be able to resolve objects 50 times fainter than those reached by the 200-inch telescope on Mt. Palomar. Free of the distortions of the earth's atmosphere, the Space Telescope will also be able to distinguish features on planets, nebulae, and galaxies ten times smaller than those seen in the best earth-based telescopic pictures. And the information-rich wavelengths of light in the ultraviolet and infrared regions, inaccessible to earthly telescopes, will fall upon the Space Telescope. Undoubtedly, astronomy and astrophysics will be revolutionized by this remarkable instrument, designed to last ten years, while the shuttle ferries back and forth its instrumental packages.

Also planned for a 1984 launch is NASA's Galileo Jupiter orbiter and probe. The recent Voyager 1 and 2 Jupiter flybys provided tantalizing peeks at the giant gas planet and its exotic moons. The Galileo orbiter will take a detailed look at the moons on repeated close passes and will monitor the swirling pastel storms of the planet. The probe will then perform a suicide mission into Jupiter's atmosphere, sampling the chemistry of its clouds.

Nearly all of the planned shuttle missions are ends in themselves, rather than means toward expanding space technology. There are some exceptions, however, which pass quietly from NASA lips, lest the agency be accused of expansionary thinking. The NASA buzz words are "*space industrialization*," a catchall phrase that applies to a range of experiments that take advantage of the weightlessness and vacuum of space. Growing perfect crystals, manufacturing precisely spherical ball bearings, depositing vapors of metals for space structures, processing pharmaceuticals and other chemicals, building large structural beams, and testing solar power systems are all modest, near-term technological building blocks we will see in the early days of the shuttle.

As we move into the late 1980s, we could see an explosion of space industries. Huge solar collectors might begin to beam solar energy to earth via microwaves to help alleviate the energy problem. A large modular space station composed of Tinkertoy-like cylinders launched aboard space shuttles could become a long-term habitat for space scientists, engineers, and physicians. A small lunar base could begin to supply raw materials that would be processed into solar collectors, structures, and fuel for a growing space enterprise. Asteroids might be retrieved to add to that supply, with masses the size of ocean liners processed into satellite power stations, permanent human habitats in orbit, large food-growing areas, and manned ships for exploration of the solar system. By the 1990s, it is conceivable that we will be well on our way toward expanding the limits to growth of a finite earth by tapping the abundant energy and materials available to us in space.

Science fiction? Probably not. Recent NASA-sponsored engineering studies show that within the next decade we can begin to industrialize space on a large scale and start to tap lunar or asteroidal resources with an investment of a few billion dollars and the equivalent of a year's worth of space-shuttle flights. Such a quantum leap is no less believable than the Apollo lunar landings would have been in 1957, before the launch of Sputnik and 12 short years prior to Apollo 11. We should also remember

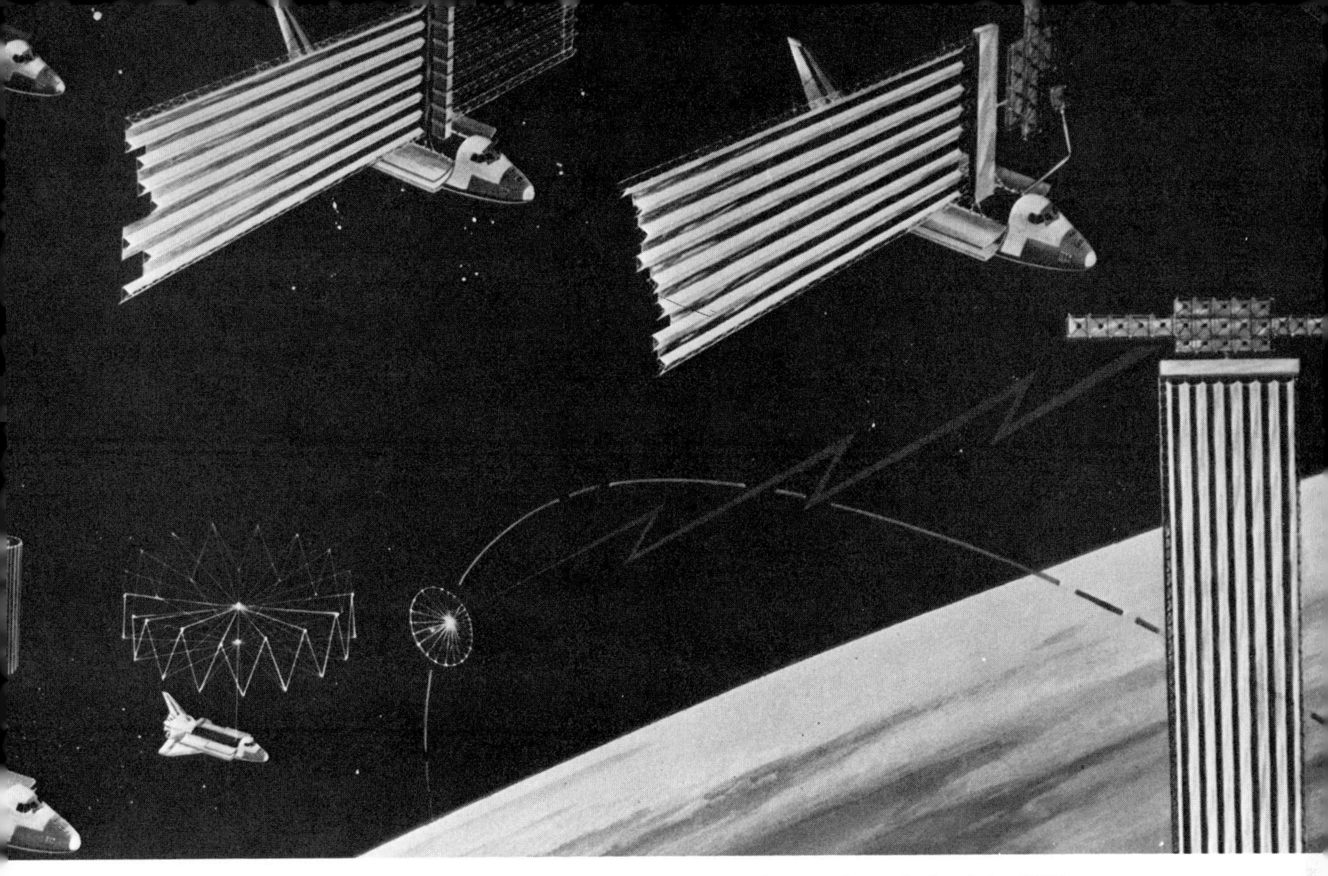

(Above) Shuttle-launched modules will comprise the space stations and test-beds of the 1980s that will pave the way to the space renaissance of the 1990s.
(Below) Small-scale sunsats like these will be tested before a full-scale commercial program.

that John F. Kennedy set the lunar goal just eight years before it was fulfilled.

"If we don't conquer space, somebody else will," said James Michener in a recent speech. The catalyst for a huge space program may again come from the Soviets, who appear to have even bigger plans than the United States shuttle. Some American analysts believe that the Soviet Union will have a booster larger than the giant Apollo-Saturn V as early as 1983 and a permanent twelve-man space station in orbit by 1985. The Russians will also have their own shuttle, a much smaller version that would ferry cosmonauts to and from earth. These plans appear to be the next logical steps beyond the recent flurry of Soviet manned flights to orbit and their stays in space of up to six months.

The strongest impetus toward an increasing worldwide presence in space is the escalating strategic arms race. The arsenal includes Soviet antisatellite killer satellites already in place and planned American and Soviet laser and particle-beam antisatellite and antiballistic missile battle stations in space. So serious are these plans that it appears probable that the United States is headed for deployment of high-energy killer lasers within the next ten years, according to a report in the July 28, 1980, issue of *Aviation Week & Space Technology*.

One study envisions the United States emplacing in orbit 25 laser battle stations with a collective force capable of destroying a salvo of 1,000 Soviet ballistic missiles shortly after their launch. Such a battery of lasers will require the assistance of more than 60 shuttle launches during the 1980s. Unless the SALT talks can catch up with military planning—and it seems highly unlikely that they will—it appears that the Americans and Soviets are headed willy-nilly toward a strategic space war that will involve the launching of enormous tonnages of weaponry into space.

While NASA's plans for the space shuttle in the early 1980s could be described as moderately interesting, we probably can expect military incentives to catapult the world into a vastly expanded human presence in space by 1990, for better or worse.

As the Cold War expands into space, we can look for another pathway to the fertile stars during the late 1980s: shuttle passenger flights. Demand for passenger service aboard a well-tested shuttle could help sustain the entire program, even NASA itself.

A passenger ticket on the shuttle would probably cost between $100,000 and $500,000, depending on the launch cost and the number of seats put inside the cargo bay. The demand could go into the thousands, since it is common for luxury ship cruises costing $100,000 or more to be fully booked well in advance.

According to *Science 80*, letters to NASA requesting passenger space on the shuttle have come from nearly 200 people, 14 of whom enclosed $500 checks to confirm their reservations. At the moment, however, NASA has no plans to offer seats to paying passengers.

An article in *Parade* magazine reported that Walter Cronkite, Zsa Zsa Gabor, Carl Sagan, and Robert Redford were hoping to be among the shuttle's first passengers. With such VIPs as drawing cards, it is conceivable that passenger service into space could become a new multibillion-dollar industry service.

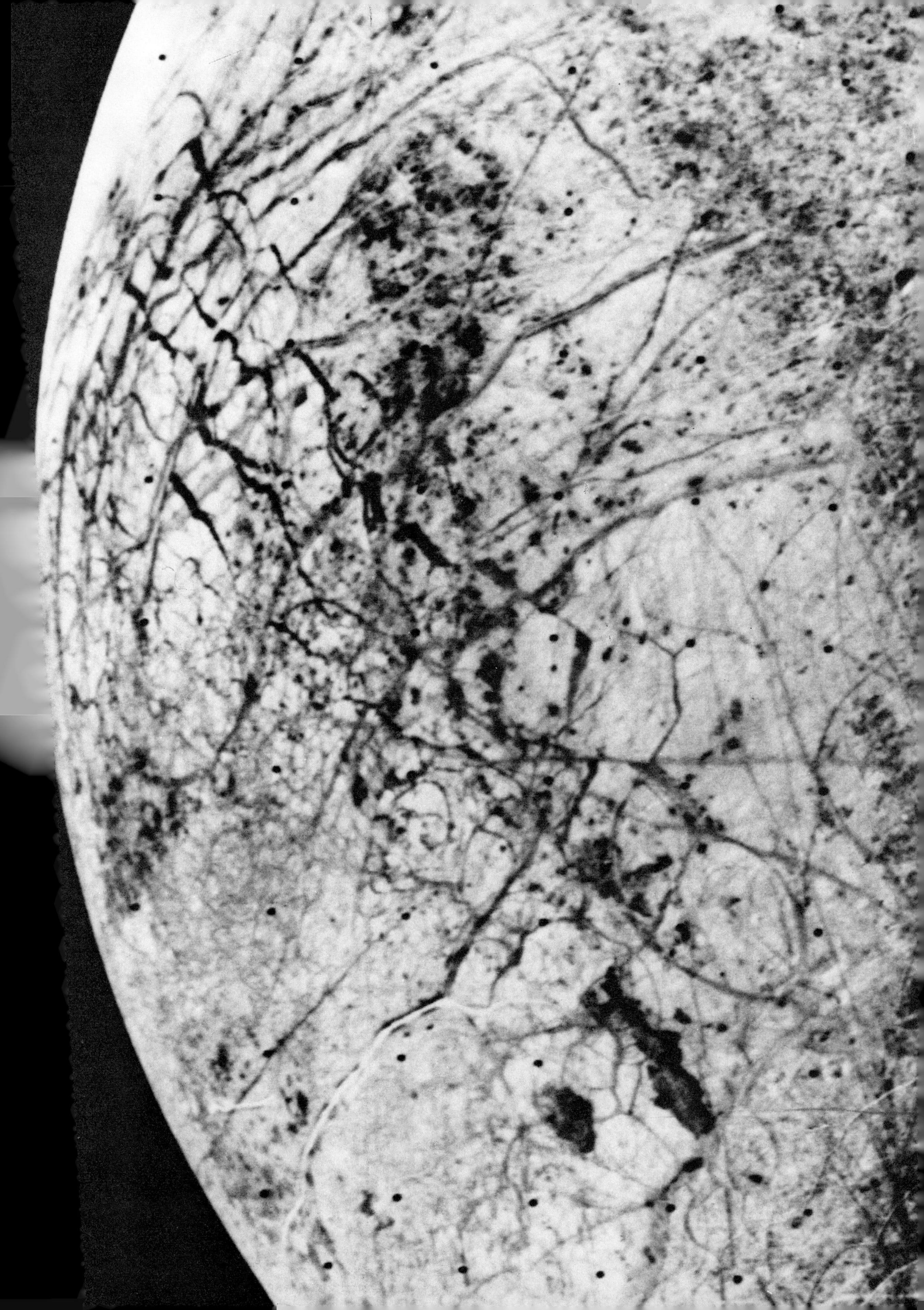

PART II

What's Out There?

Cracks and crevices crisscross an otherwise smooth Europa, Jupiter's second Galilean satellite.

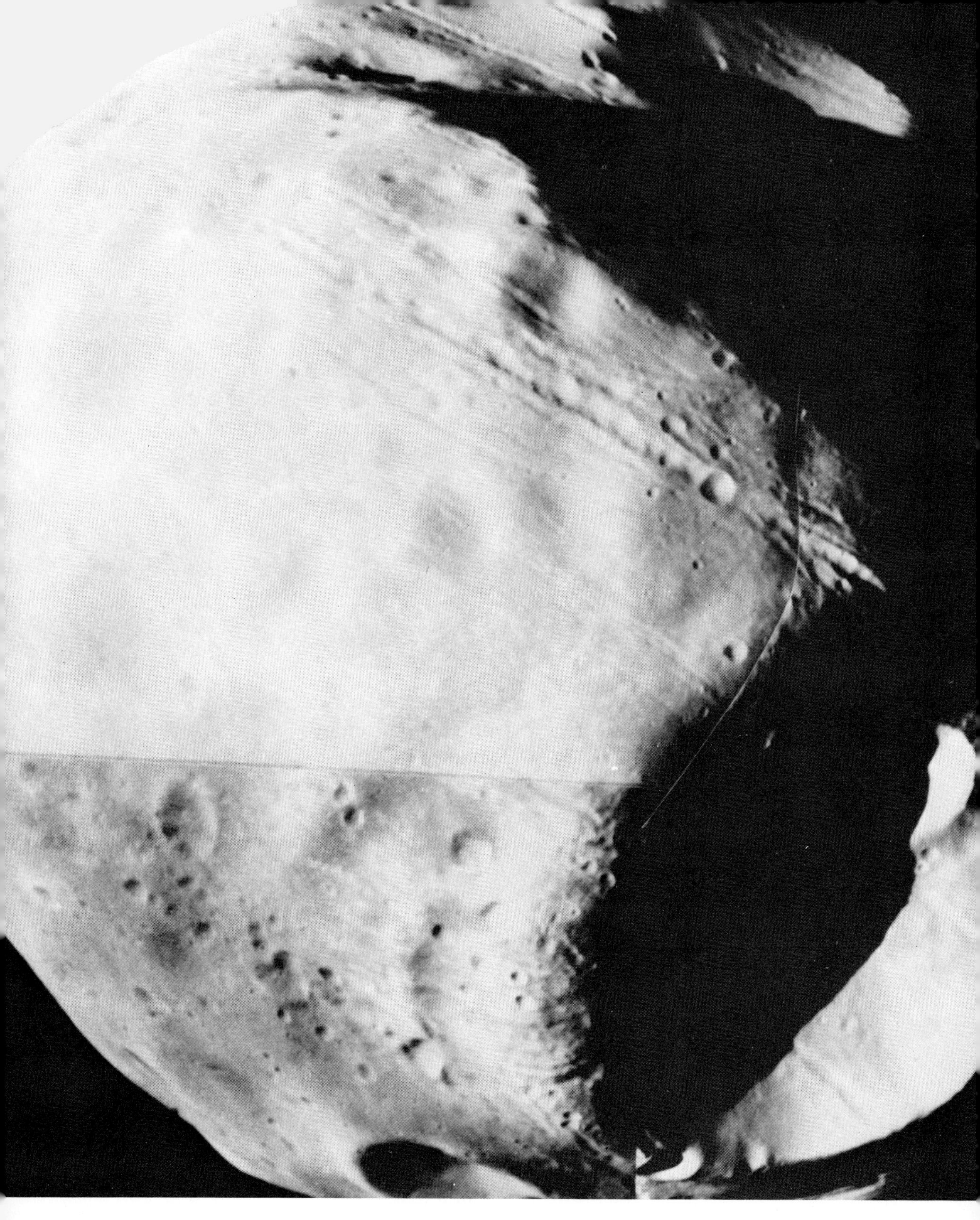

The Martian moon Phobos is an accessible fertile star.

3. A Piece of the Rock

Phobos is an odd, rocky, potato-shaped moon of Mars measuring about 20 kilometers at its widest point. In 1976, the U.S. Viking orbiter photographed Phobos, and the picture showed this lumpy moon to be pocked with thousands of craters from the impact of meteoroids over the eons. The smallest discernible crater in the picture measures barely 100 meters across, about the size of a football field. Yet a 100-meter chunk of Phobos is adequate material to build enough satellite solar-power stations in earth orbit to meet increased American electricity demand between now and the year 2000. Phobos is a fertile "star." (A star in the broadest of context, as the ancients would have described planets, asteroids, and moons, as well as the gaseous stars themselves)

The study of Phobos has had an interesting history, and its ultimate fate may be even more interesting. Using mathematical reasoning that had no relation to physical reality, the famous German mathematician and astronomer Johannes Kepler speculated in 1610 that Mars had two moons. In the following century, the literature of Swift and Voltaire mentioned the two moons, yet there was still no observational evidence that any such bodies existed.

Not until 1877 did the moons of Mars materialize in telescopic observation. At the U.S. Naval Observatory, Asaph Hall noticed two faint, starlike specks moving in orbits around Mars. He named them Phobos and Deimos, after the two sons of the Greek god Ares (Mars). (*Phobos* is translated *fear*; *Deimos* means *terror*.) Kepler's unscientific guess proved to be correct, a puzzling historical event attributed to coincidence.

Unlike the earth's moon and the four large satellites of Jupiter discovered by Galileo with his small telescope nearly four centuries ago, Phobos and Deimos are tiny moons, barely observable through large modern telescopes. In fact, both objects resemble asteroids more than they do the large, spherical moons and planets. Asteroids (*aster* is the Greek word for *star*) are starlight-speck objects that occupy the large reaches of space between the orbits of Mars and Jupiter. The thousands of known asteroids range in size from the faintest dots of light seen moving across the fields of the giant telescopes to the largest object in the asteroid belt, Ceres, about 1,070 kilometers in diameter.

The pull of the planets sometimes sends asteroids into new orbits that cross the orbit of Mars and, occasionally, even the earth's. These asteroids form a special class of objects called the Apollo (earth-crossing) and Amor (earth-approaching but not crossing) asteroids. We know of the orbits of about fifty Apollo and Amor objects, and the discovery rate is increasing all the time. Most Apollo asteroids will crash into the Earth or Venus within the next ten million years. Because of their orbits and the frequency of meteor observations, some of the Apollo objects are believed to be the parent bodies of many of the meteorites that have struck the earth. Like Phobos, a large number of Apollo and Amor asteroids are fertile stars: they are easy to reach and they are made of materials that will become useful to us in space.

Whether Phobos and Deimos are Mars-captured asteroids or were formed out of dust when Mars accreted some four to five billion years ago, we do not yet know. The debate hinges on finding out the chemical composition of these rocks. Some scientists favor the viewpoint that they were not captured because of the difficulty of explaining their almost perfectly circular orbits above the Martian equator. Furthermore, their very dark colors suggest a "carbonaceous" composition, like some of the meteorites that land on the earth. If that is so, then Phobos will probably become a very fertile star indeed, chock full of water (bound in rock crystal, up to 20 percent of the total weight) and carbon (several percent by weight).

Observations of the orbit of Phobos suggest that it is very gradually speeding up. The data led Soviet scientist I. S. Shklovskii to suggest in 1960 that Phobos was a large, hollow artificial satellite with a decaying orbit, dragging along the outer fringe of the atmosphere of Mars. Calculations showed that Phobos would meet the same fate as Skylab 1 and crash into Mars. (Skylab 1 was an ill-conceived mission, where the designers didn't make the orbit high enough.) More modern observations, verified conclusively by the natural appearance on spacecraft photographs of Phobos and Deimos, show that Phobos is indeed accelerating, but that a hollow object skipping and dragging along the atmosphere was not a necessary or appropriate explanation. Tidal drag was an adequate explanation, tidal drag being the differential pull of Martian gravity on the front and back of Phobos, causing it to lose some energy and fall in toward Mars. At its present rate of fall, Phobos should crash onto the Martian surface in about 100 million years.

The specter of fear raised by the collision of a fair-sized asteroid with a civilized planet (and Mars may become inhabited by then) was dramatized in the movie *Meteor*, in which a Phobos-sized errant asteroid called Orpheus threatened to crash onto earth. A few small pieces of Orpheus (one of the largest of which cratered Central Park in New York City) created enough havoc to produce the most devastating disaster in human history. More on hazards later. Suffice it to say that the fertile stars share the trait of being a potential source of disaster, as well as being the potential source of abundant energy, food, and raw materials.

Phobos may be one of our first stops in the quest for new resources. Brian von Herzen (a Princeton physics student) and I calculated that the total amount of energy (and therefore the cost) required to go from the earth to Phobos, take a piece of it, and return it to earth orbit for processing into satellite power stations, space stations, and other products, is less than that for any other known object in the solar system—including our own moon!

The key is energy and gravity. The velocity of escape from Phobos (15 meters per second) is 150 times less than that from the earth's moon. Because the required energy is proportional to the square of the velocity, it takes almost no effort or fuel to land on Phobos and escape from its gravitational pull, unlike the moon.

At the same time, we know our moon is a fertile star. The energy needed to lift material from the lunar surface is one-twentieth of that required from the deep gravity well of the earth. The moon contains an abundance of oxygen, silicon, and metals, all of which will be useful to us in space. As we shall see, the moon may become a significant stepping-stone to Phobos and the asteroids.

It is possible to envision a mining operation on Phobos in the year 2000. An important scientific dividend of such an operation would be the use of Phobos as a base for the exploration of Mars.

Phobos—this odd-shaped rock, rich in legend and history, and perhaps doomed to extinction within a small fraction of the future lifetimes of the sun and planets—may soon become the focus for the exploration and exploitation of the solar system.

4. *The Making of a Concept*

In the summer of 1969, when Neil Armstrong first set foot on the moon and the war in Vietnam was raging, I had a visit from Princeton professor of physics Gerard K. O'Neill in Ithaca, New York, where I was teaching at Cornell. O'Neill and I had originally met as roommates in the tryouts for the astronaut program two years earlier. (I had since been selected, become an astronaut, and then "unbecome" an astronaut—a saga presented in my earlier book *The Making of an Ex-Astronaut.*) A common characteristic of the week-long, world's-most-thorough physical and mental astronaut-candidate examination was that roommates became friends for life, as if they were military combat buddies. We both felt that way after a

In this photograph taken in 1967, astronaut candidates O'Neill and O'Leary grin after the world's most thorough medical examination.

rigorous week at the U.S. Air Force School of Aerospace Medicine in San Antonio, Texas.

At the time of O'Neill's visit in 1969, there was a great deal of unrest and an increasing lack of confidence among Americans, particularly on college campuses. People were beginning to realize that the world would eventually run out of resources and would experience growing limitations on energy use.

O'Neill, with a glint in his eye, said he thought he might have a solution. O'Neill believed that it would be possible within today's technology for man to permanently settle in space without spending much money on the initial investment, perhaps something on the order of the Apollo program, or one percent of the federal budget for ten years. Once you establish this first beachhead in space, you can expand *ad infinitum*, because solar energy is available in space all the time and you have materials available to build large structures for living in space that are easily retrievable from very shallow gravity wells.

In layman's language, it is technically feasible to mine the moon or an asteroid and gently nudge the materials to a manufacturing site in space where continuous solar energy enables you to process the materials into large structures, like a space settlement housing 10,000 people, rather than spending billions of dollars to launch the same materials from earth. You just push things around a little, and in time the combination of materials and energy will create a multitrillion-dollar earth-independent economy.

One of the basic principles of business is that wealth is generated by three things: materials, energy, and intelligence. What seemed most interesting to me about what O'Neill said 12 years ago was that we had all three elements in enormous quantities. The major obstacle was communicating this to other people, because there is a gestation period for accepting such an unusual idea, even if it could be accomplished within our own lifetime. In the decade since our discussion, there has been an exponential growth in our understanding of and the popularity of this concept.

How would it feel to be inside a 10,000-person space habitat? It would be made of metals, glass, and soil mined on the moon or an asteroid, but we wouldn't know that the materials were unearthly once they were processed in space. A colony of this size could well exist with a small investment, compared to the amount of money we are spending here on earth. Before the year 2000, a space colony could become a fertile star, containing villages, trees, grass, animals, ponds, and hills. It need not be the quasi-military, antiseptic, battleship-gray product seen in science fiction movies. The space-cabin environments envisioned in *Star Wars*, *Star Trek*, and *Buck Rogers* are unnecessarily archaic.

One design for an early space habitat calls for a sphere about one mile in circumference that spins twice a minute, creating an artificial gravity on the inside wall so that a person standing at the equator experiences one earth gravity, just as if he were on earth. As one walks toward a pole, he gets lighter and lighter until, at the pole itself, he can float.

Building space colonies is not a new idea. At the beginning of this century, the Russian schoolmaster and physicist Konstantin Tsiolkovsky

O'Neill suggested that large space habitats like this one could be built from lower materials.

envisioned huge greenhouses in space to provide food for the colonists. The vast source of materials that made all this possible was the asteroids.

This theme was echoed in various forms by numerous others. At the beginning of the space age, for example, Dandridge Cole and some of his colleagues wrote about asteroid colonies. It was not until the 1970s, though, that the various ideas all came together in a confident, hard-headed economic and engineering sense. Gerard O'Neill was principally responsible for the grand synthesis: the development of a plausible scenario *using current-day technology* for the mining, transport, and processing of lunar materials into a largely earth-independent, geometrically expanding industrial base in space. Finally, between 1975 and 1980, people began to really listen.

5. *The Fertile Stars*

People have begun to listen seriously to speculation about space colonies because they are becoming increasingly aware that the earth is a very fragile, finite planet. There are more than four billion people on the earth right now, and in the year 2000 there will be six billion of us. And there's nothing we can do about it, even through increased population control.

The limited energy resources that are available to four billion people put us on a collision course with disaster. We are running out of oil, with only enough to supply us for another one or two generations. Food and raw materials are also becoming increasingly scarce. Massive shortages are projected for the twenty-first century, when human survival may be threatened by overpopulation, overexploitation, pollution, starvation, and psychological limitations to freedom and exploration.

Because of earth's gravity, it would be very expensive to send materials into space. Using the space shuttle planned for the 1980s, it would take millions of tons of material, at $1,000 a kilogram, to create habitats and power satellites. A solution is to utilize materials from the moon.

Starting in 1969, the Apollo program gave us a boost in confidence concerning the use of lunar materials. The samples brought back from the moon by the astronauts were rich in oxygen (about 40 percent), metals (20 to 30 percent), and silicon (about 20 percent). These three substances would be useful to us in space, provided we could mine them, carry them into space, and process them into usable products. Oxygen can be used for breathing and rocket fuel, metals for space structures, and silicon for glass and solar collectors.

O'Neill pointed out that the energy required to lift materials from the moon into space is about one-twentieth of that required to launch materials from the earth, because the moon has a much weaker gravity. Launching objects from the moon is like chucking something out of a well that's one-twentieth as deep as the well that you would start from on earth. The moon is more of a gravity dimple than a gravity well. The major reason we haven't been able to develop human enterprise in space is that on earth we are sitting at the bottom of a very deep gravity well.

We live in a time when we are compelled to take a new look at our surroundings, a time not unlike the beginning of the Renaissance, when Nicolaus Copernicus, a Polish physician, canon, and astronomer, overturned the prevailing notion, held since the time of the Greeks, that the earth occupied a special place at the center of the solar system and the universe. Putting the sun at the center of the solar system made more mathematical sense, but the thought was humbling to the mind of man, and it took decades to be fully accepted. It is comfortable for us now to say that pre-Copernican thinking represented a form of planetary chauvinism.

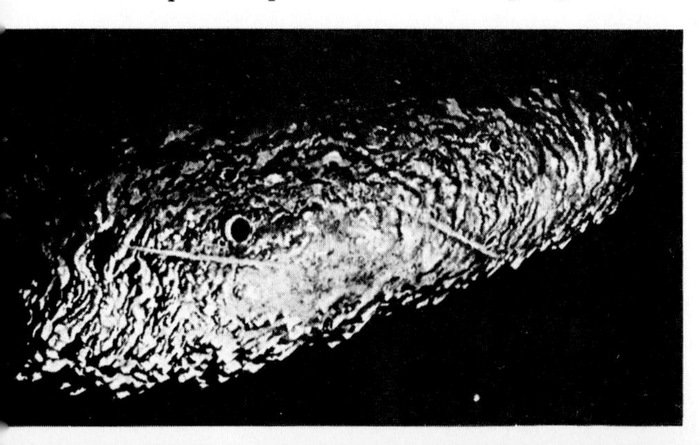

(Left) Artist's concept of the Earth-approaching asteroid Eros, which will be easy to reach during the early 1990s.

(Below) The Moon contains silicon, oxygen, and metals that will be useful to us in space.

We are on the verge of a second Copernican revolution, this one concerning a gravitational chauvinism. Freeing ourselves from bondage of earth's gravity may be no less profound a leap in the perception of ourselves and our planet than the original Copernican revolution was nearly 500 years ago.

The currently accepted, but mistaken, belief is that you have to launch *everything* from the earth in order to do *anything* in space. But the new perception is to launch from the earth only what you need in order to mine the moon or the asteroids and take from them the materials you require to build things in space, where you have constant solar energy. By means of this bootstrapping procedure, we can create and produce in the vast field outside the earth's gravity well.

Three years ago, I suggested that some of the earth-approaching (Amor) asteroids are the most accessible and versatile source of materials in space. These rocks weigh millions to trillions of tons, and there are a lot of them. If we could bring a tiny asteroid—the size of a football field—near the earth and mine it in space, we would acquire a million tons of material, enough to build a space habitat for 10,000 people.

It turns out that there are about 200,000 asteroids weighing more than a million tons each that pass close to the earth at some time. There are times when many of these objects are within reach of earth's telescopes. We now know the orbits of fifty Apollo and Amor asteroids; that number is likely to reach several hundred by 1990.

The impetus for the asteroid-search program comes from the research of Eugene Shoemaker, Eleanor Helin, and their colleagues at the California Institute of Technology. Using what Shoemaker describes as "practically nineteenth-century science" on a shoestring NASA budget, these scientists have discovered a dozen earth-approaching asteroids. Their diameters range from a mere 100 meters to the irregular 10-20 kilometers of Eros. They occupy orbits around the sun that either intersect or closely approach the earth's.

During their monthly trek to Mt. Palomar, Helin and her students spend four or five nights photographing small patches of sky with an 18-inch Schmidt telescope. During the twenty minutes or so that it takes to obtain one picture, an occasional earth-approaching asteroid betrays itself as a faint streak against a stationary background of planets, stars, galaxies, nebulae, and far-away asteroids. The fast apparent motion of this asteroid results from its closeness to the earth. (It's much like taking a picture from a moving car when an unwelcome telephone pole blurs the foreground.)

Helin's work has barely begun when she finds such a streak. She must determine whether it is really an asteroid or one of hundreds of artificial satellites orbiting the earth. A second picture supports (or disappointingly denies) the finding, but the work has still not ended. The object needs to be tracked for several days or weeks to confirm its existence and to establish its orbit. This usually requires alerting astronomers at other telescopes. Otherwise, it will probably rapidly fade in brightness and disappear into permanent oblivion.

As in most fishing expeditions, the one-that-got-away happens all too often. Persistence and a lot of hard work have made Helin's efforts pay off. After the discovery of 1976AA and 1976UA (two asteroids with orbits very similar to the earth's), 1977HB (a good candidate for a mission), and many more after that, the bandwagon is mounting.

Asteroids that sometimes pass inside the earth's orbit are called Apollos, after the name given to the first one discovered in 1932 by Karl Reinmuth at the University of Heidelberg. He saw the streak of Apollo in the course of a photographic search for ordinary asteroids.

Since the naming of *Apollo* nearly 50 years ago, the *earth-approachers* have been given such colorful names as Hermes, Eros, Anteros, Belulia, Geographos, Toro, Icarus, Bacchus and, most recently, Ra Shalom, which commemorates the recent Israeli-Egyptian Camp David accord. *Ra* is the name of the Egyptian sun god, and *shalom* is the Hebrew word for *peace.*

Reinmuth correctly identified Apollo as an earth-crosser, but unfortunately it was lost because it was not observed long enough to obtain its orbit. By a remarkable coincidence, Richard McCroskey and Cheng-Yuan Shao, at Harvard College Observatory, rediscovered Apollo 41 years later in the first photograph they took in what was to have been a several-month search. Good fortune occasionally befalls the Apollo-fisher.

Palomar astronomer Charles Kowal, a regular observer at one of the world's most sophisticated and sensitive survey telescopes, has caught Apollo fever. Having spent most of his career surveying objects outside our galaxy at the 48-inch Schmidt telescope, he now examines his large, sharp, star-filled plates for narrow, faint streaks and blobs. As a result of his new interest, Kowal discovered the unique mini-planet Chiron, which inhabits the darkness between the planets Saturn and Uranus.

Kowal has also discovered several earth-approachers and now welcomes Helin and her co-workers at the telescope to try their luck. Meanwhile, observers in West Germany, Japan, Chile, and Australia are now playing the Apollo game, so Helin has become a world traveler. "It's tiring," she recently said to me, "but well worth the effort."

Shoemaker and Helin hope to obtain a 48-inch Schmidt telescope dedicated to augmenting the search for earth-approaching asteroids. This wish is backed by the unanimous recommendations of several NASA working-group panels, but its fulfillment awaits the lifting of the austere NASA budgets for 1980 and 1981. Such an instrument would permit the discovery of several hundred new earth-approachers, many of which would be prime targets for mission. The Air Force also has sophisticated instruments that could increase the discovery rate, if they could loosen a bit their tight lid on security.

We have barely begun to identify the asteroids in the solar system. Shoemaker and others estimate that, over time, 100,000 asteroids with diameters greater than 100 meters pass close to the earth, many of which could be detected by existing telescopes.

There is another revolution going on in asteroid science. In recent years, astronomers Thomas McCord, Michael Gaffey, David Morrison, Dennis Matson, Lawrence Labofsky, Carle Pieters, Glenn Veeder, Joseph Veverlea, Clark Chapman, Ben Zellner, and others have determined the chemical compositions of several hundred asteroids by observing how they reflect sunlight. In some cases, they can make specific identifications of metals, minerals, and water.

The data strongly suggest that the asteroids fall into the same classes as meteorites: some are stony, consisting mainly of silicon and oxygen; some are metal-rich (in extreme cases nearly 100 percent iron and nickel); and others contain significant amounts of water and carbon.

The rich variety of chemical compositions and orbital distributions provide clues about the origin and evolution of the solar system. The asteroids appear to be fragments of larger planetary bodies accreted early in the history of the solar system. They have apparently not undergone as many chemical and thermal changes as the planets and large satellites. Therefore, asteroids may provide the most direct clues about the early state of the solar system.

Understanding the origins of the earth-approaching asteroids is a complex business. Shoemaker and his colleague George Wetherill of the Carnegie Institution of Washington, D.C., suggest that some of the earth-approaching asteroids are the cores of defunct comets that originated in the outer fringes of the solar system, while others are asteroids perturbed inward toward the sun by the planets from the main asteroid belt between the orbits of Mars and Jupiter. In both cases, these bodies, if directly sampled, should provide a treasure trove of knowledge about what was happening early in the formation of the solar system at varying distances from the sun.

The earth-approaching asteroids and comets appear to be the parent bodies of meteorites. Analysis of meteorites reveals a wide variety of chemical types and histories, some going back to the primordial solar system. Unfortunately the samples are not completely pristine. Meteorites are weathered by their fiery entries into the atmosphere and unrelenting water erosion before they are picked up and brought into the laboratory.

By looking at the distribution of sizes and the freshness of the larger craters on the moon and inner planets, we can get an idea of the history of impact by earth-approaching bodies. This is harder to do on the earth because all but the most recent craters have been obliterated by water and weathering.

Counting lunar, Martian, and Mercurian craters reveals that during the early epochs of the solar system there were a lot more Apollo-type objects floating around than there are now. Most of the objects were swept up by the planets, but the asteroids inhabiting the main belt between Mars and Jupiter had stable enough positions to survive the eons of uneven planetary pulling. Still, it can be shown that those of them in orbits toward the inner part of the belt can be driven into the inner solar system to help replenish the supply of earth-approachers. The rate of replenishment appears to be insufficient to account for the entire supply of Apollos, however, so we are left with the scientific puzzle: where do most of the Apollo asteroids come from?

On the basis of meteorite analysis and telescopic observations of the light reflected from asteroids, Wetherill suggests that most iron meteorites and a significant number of stony meteorites come from the inner part of the main belt. He adds, "Because Apollo objects are in orbits which come close to the earth, and also must be fragmented as they traverse the asteroid belt near aphelion (the farthest point from the sun), there must be a component of the meteorite flux derived from Apollo objects." In other words, most, if not all, meteorites are either Apollo objects or the remnants of collisions between Apollo asteroids and main-belt asteroids. But the origins of the Apollo objects themselves are likely to be dead comets *and* errant main-belt asteroids.

In any case, we need a source of replenishment for earth-approaching asteroids because they cannot survive in their current unstable orbits. Scientists who study the changes in their orbits have concluded that all currently observed earth-approaching asteroids will eventually impact the earth, Venus, or the moon, and most of the events will occur within the next 100 million years, a mere fiftieth of the age of the solar system. On this time scale, Apollo asteroids are short-lived objects. On the scale of a human lifetime, the hazard of an earthly cataclysm is remote.

Some of the earth-approaching asteroids are remarkably accessible to spacecraft that venture beyond the earth's gravity. Physicist John Niehoff has calculated that the launch of one space shuttle in 1992 with an upper-stage rocket housed in its cargo bay is all that is required to loop in a lazy coast around the sun to rendezvous with the asteroid Anteros, grab a one-kilogram sample, and return it to the earth's surface three years later.

Small asteroids have next to no gravity. Like the moons of Mars, Deimos and Phobos, their shapes are irregular, as shown by variations in their brightnesses as they rotate. Although asteroids have never been photographed at resolutions greater than starlike points, we can be certain they will appear as rocky lumps, like Phobos and Deimos, rather than spherical, because their gravities are not sufficient for the minerals to overcome their own bonding strength and crush into spheres.

A person standing on a three-kilometer diameter asteroid could easily jump off its surface and never come back. The lack of gravity turns out to be a convenience to the visitor: no retrorockets are needed to "soft-land" a spacecraft. Even the low-gravity moon requires as much fuel for a soft-landing as the amount of payload that is actually landed. Taking off from the moon would consume even more fuel.

The energy cost of going to a convenient earth-approaching asteroid can sometimes be less than that required to land on the moon. To land and take off from the moon, we will probably need high-impulse chemical rockets that can do the job rapidly.

Such is not the case for the asteroids. It is possible to gradually alter an orbit by coasting. This can best be done by means of a solar-powered propulsion system like those now being developed by NASA. One idea is to use an electromagnetic motor called a *mass-driver.* Solar-powered superconducting magnets lining a long, narrow tube accelerate pulverized asteroidal material as reaction mass, or fuel. In this way, very large asteroidal fragments can be moved cheaply; the energy (from the sun) and fuel (from the asteroid) are free.

The sun shines all the time in space away from a planetary surface. Solar collectors near the sunny side of an asteroid could supply a continuous source of power that could maintain a scientific station, propulsion system, and mining operation. On the other hand, any spot on the earth, moon, or planets suffers the intermittency of daylight, the need for solar collectors to track the sun while fighting gravity, and, in some cases, the vicissitudes of weather.

The vision of earth-approaching asteroids as abundant, accessible, cost-effective, and versatile sources of materials for space industrialization is not new. As early as 1900, the Russian scientist Konstantin Tsiolkovsky envisioned mankind migrating to and inhabiting the asteroids. This vision has since been amplified by Dandridge Cole, Arthur C. Clarke, Gerard K. O'Neill, and numerous others.

Recent progress in asteroid science, the existence of the space shuttle, the development of satellite solar power and space manufacturing of nonterrestrial materials as feasible engineering concepts, and our perception that mankind is rapidly depleting the earth's energy, food, and metals, have suddenly elevated the study of asteroidal resources from the realm of science fiction to technological and economic reality.

6. *Power from Space*

Around the time Gerard O'Neill was first propounding his ideas about space colonies, Dr. Peter Glaser at the Arthur D. Little Company was just beginning to promulgate a concept almost as radical: placing huge solar collectors in space, beaming the energy down to earth by microwaves (radio waves), and converting the energy to electricity for residential and commercial use.

These "sunsats" would occupy 24-hour, or geosynchronous, orbits 23,000 miles above the equator. To an observer on earth, they would hang stationary, as if tied by a string, keeping the same position in the sky while the sun, moon, planets, and stars make their apparent daily circuits across the sky. (Communications around the earth are now being revolutionized by Comsats and Intelsats, which also sit in geosynchronous orbits.) Each sunsat would measure several kilometers across and be capable of harvesting billions of watts of energy, the equivalent of several modern fossil-fuel or nuclear-power plants that could power a city the size of Los Angeles. The receiving antennas, or rectennas, would form a circular array several kilometers across. Some day sunsat systems may replace coal, oil, and nuclear-power plants as the mainstay of our continuous (baseload) central-station electricity production. The earth's existing power grid system could be tapped into sunsat rectennas.

The sun shines virtually all the time on objects in geosynchronous orbits. Sunsats would enter the earth's shadow only a tiny fraction of the time, a few minutes near local midnight during the equinoxes in March and September (when candlelit blackout parties could be planned), removing the major objection—intermittency—to most forms of earth-based solar power, which currently require expensive energy-storage systems for nighttime and cloudy periods.

Each satellite would lock into its own rectenna, which could be located anywhere on earth except near the poles. The microwaves would be low-density, less than the intensity of direct sunlight, and the amount of energy received just outside the fence surrounding the receiving antenna would be several hundred times less than U.S. tolerance standards for exposure to

Automatic beam-builders spin their webs of sunsat structures.

microwaves. The receiving antennas would permit agriculture beneath their wire mesh. A given area of land would receive solar energy about 35 times more efficiently and put less waste heat into the atmosphere than earth-based solar collectors.

An increasing number of scientists, engineers, and economists believe that sunsats may be the ultimate solution to the global energy problem, that they are an economically competitive and environmentally compatible source of continuous solar power, which could replace nuclear and coal power by the turn of the century.

Sunsats transmit energy efficiently through the earth's atmosphere and they are technically feasible. In a recent test, an antenna at the Goldstone facility in California beamed microwaves to a collector a mile away, lighting a set of headlights. This is experimental proof that energy can be efficiently transmitted through the atmosphere by microwaves and be converted into electricity. Laboratory experiments show that the overall efficiency, from collected electrical current to final current delivered from the antennas, is 60 to 70 percent, with projections of 80 percent or higher for the transmission from satellites in space.

The alternatives for providing continuous electricity are bleak. As uranium supplies run low, nuclear power would eventually have to depend upon breeder reactors, which involve the production of large quantities of deadly plutonium, the material of atomic bombs. The problems of radioactive-waste disposal and reactor safety have not been dealt with adequately, and nuclear power may not be economically or politically acceptable. Since the Three Mile Island accident, there has been a virtual moratorium on new nuclear power plant construction.

Fossil-fuel plants will need to depend on coal, because we are running out of oil, but coal is dirty and the carbon dioxide emitted by the power plants may place an intolerable heat burden on the earth's fragile biosphere. Other forms of solar power are intermittent and still too expensive. Nuclear fusion has not yet been proven feasible and has its own environmental and economic drawbacks.

And yet the need is there: utilities are planning to invest several hundred billion dollars on new capital equipment for generating electricity in the United States over the next two decades, and we continue to import $50 billion worth of foreign oil every year. Sunsats possess all of the advantages of earth-based solar power—they are clean and no fuel is needed—and none of the disadvantages—expense and intermittency.

So, we ask, why not develop satellite power? Three problems have been identified: the high cost of launching these huge satellites (the size of the ship Queen Elizabeth II) into geosynchronous orbits; the environmental impact of the launch fuel that would be spewed into the atmosphere; and the environmental impact of the transmission of microwaves through the atmosphere.

Launching materials from the earth for one satellite power station would require 100,000 tons of material and cost $100 billion using the space shuttle. Launching, rather than building the station, would be the biggest cost because of having to escape the earth's gravity well. The cost of

This array of earthbound antennas will receive and transform enough electricity to power the city of New York.

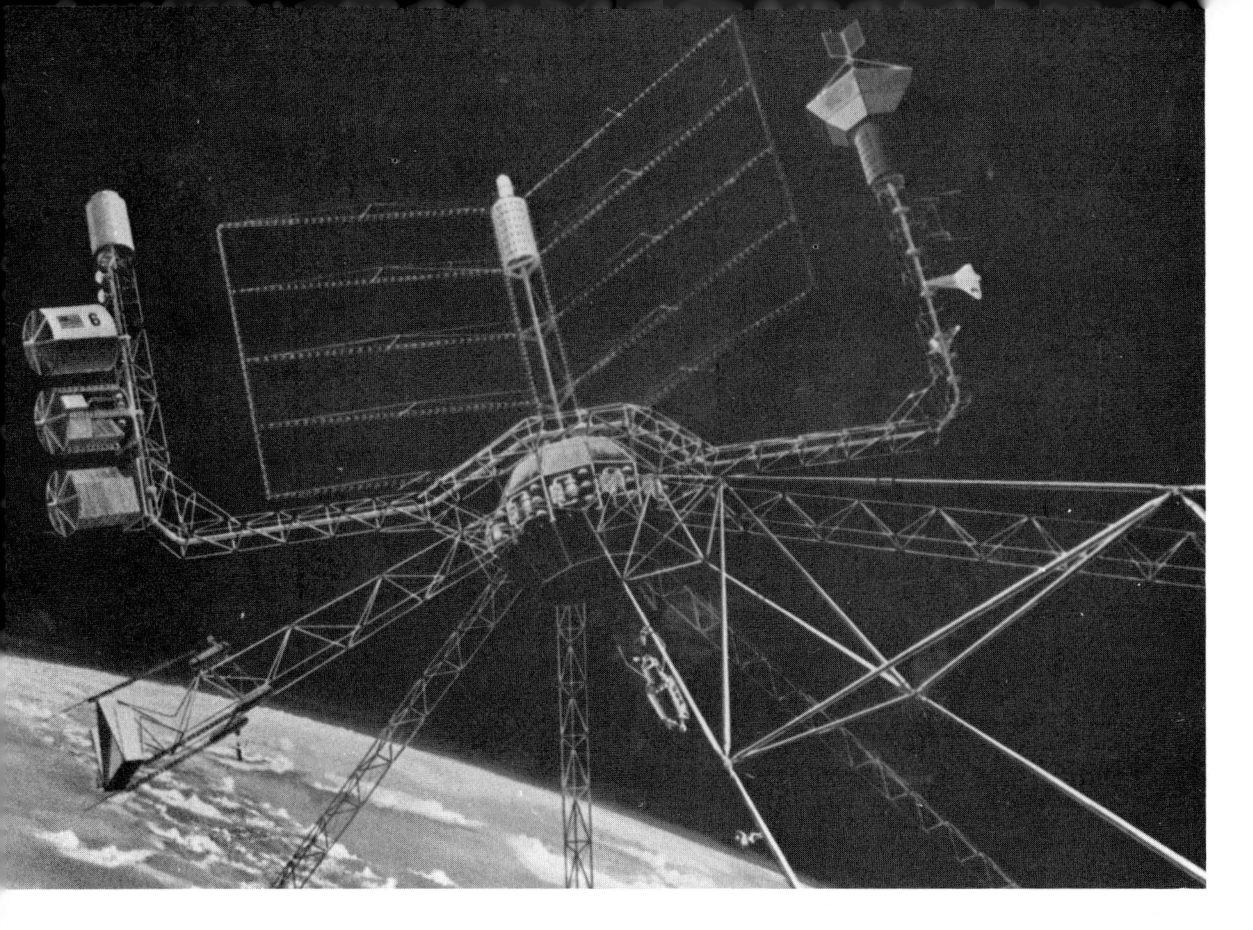

(Above) One sunsat design calls for a generator that would run from heated sunlight.
(Below) This space factory could produce the first sunsats as early as 1990.

launching a full fleet of sunsats by space shuttle would exceed the U.S. gross national product! Furthermore, you'd have to launch a shuttle flight every few minutes carrying the materials necessary to construct a satellite power station large enough to make an appreciable dent in worldwide energy requirements. The cost would be prohibitive.

To help remedy the problem, NASA has on the drawing boards a heavy-lift launch vehicle, which could carry several hundred tons of material on each flight. But some studies have shown that the effluence from these launches—more than a thousand times greater than that of the space shuttle over time—could produce as much damage to the stratosphere as a fleet of supersonic transports. The result could be the depletion of ozone, which protects us from the deadly ultraviolet radiation of the sun.

Adverse effects on other parts of the atmosphere have also been identified. Even if these behemoth rockets were to pass environmental tests, the launch costs still seem to be too expensive to bring satellite power into competition with other sources.

This is where space manufacturing comes in. In 1975, O'Neill married his ideas with Glaser's concept. He asked why the sunsats couldn't be made from lunar materials. The space shuttle, perhaps with some mild upgrading ten years from now, is all that is necessary to mine the moon and asteroids and process those materials in large quantities into satellite power stations and other products in space. Engineering studies have shown that the initial investment is minuscule compared to the total return from the program. They concluded that at least 96 percent of the power satellites could be built from lunar silicon and metals. Hundreds of tons of materials initially launched from the earth could give leverage on up to several hundred thousand tons of structures manufactured in space. The space shuttle has passed environmental scrutiny; anything much bigger might not. Maybe nature is telling us something.

Two of the three objections to satellite power appear to be answered by going the extraterrestrial route. The economics, expressed in the cost of electricity delivered to the consumer, look competitive with existing power sources *at current prices;* they will probably look much better in the year 2000.

The answer to the third objection, the environmental impact of microwaves, awaits further testing. Studies funded by the U.S. Department of Energy have found no show-stoppers. There is considerable disagreement in the environmental movement about the acceptability of satellite power. At times the debate is shrill, reminiscent of the debate on nuclear power. The area of greatest disagreement concerns what the influence of very low dosages of microwaves might be on human beings outside the fence and on birds that fly or roost within the beam. The fact remains that a system can be designed to reduce the exposure to humans considerably below current tolerance standards, lower than average daily exposures to microwave ovens or to radio transmitters atop New York City skyscrapers. Even at the center of the beam, the amount of heat a bird would feel would be half that of direct sunlight. Airplanes, with their metal shields, could fly through the beam with no harm to passengers.

The beam cannot wander. Its very existence depends on a command sent up from the receiving antenna. When the "phase lock" is broken, the beam turns off and radiates diffusely and harmlessly into space.

The beam would make a poor weapon, even if retargeted toward an enemy capital. Because of the sunsat's 23,000-mile distance, the laws of physics tell us that the density of the radiation cannot be greater than about one-half the intensity of direct sunlight. This quantity is sufficiently low that a potential victim would have time to shield himself from this low exposure, less than what most radar technicians and occupants of offices high in skyscrapers experience every day.

The term "microwave radiation" tingles the spine and has been misinterpreted by some environmentalists. It is often confused with the ionizing radiation caused by radioactive materials in a nuclear bomb or power plant. The high-energy ionizing radiation of nuclear reactions strips atoms in human and animal tissues, alters their chemistry, and can lead to leukemia and other forms of cancer. Microwaves can also penetrate into the skin. But the atoms and molecules only heat up and vibrate; there are no chemical changes. Overexposure *can* cause cataracts, but there is no evidence that fatal diseases result from even fairly heavy exposure to microwaves. In any case, the sunsat beam density can be made acceptably low, and, on paper at least, it appears to be winning the environmental battle with nuclear and fossil-fuel power.

Some people object to the potential military vulnerability of sunsats, but in our age all power sources, whether on earth or in space, are vulnerable to attack. Actually, because of their great distance, it may be easier to defend a sunsat than a target on earth.

Some environmentalists contend that microwave beams will interfere with radio communications and radio astronomy conducted on earth. These problems could be removed by using communication satellites, rather than reflections off the earth's ionosphere, and by moving radio telescopes into space. Both projects will probably be taking place over the coming decades anyway, quite apart from the development of sunsats.

Others argue that microwaves will heat the earth's atmosphere and raise its mean temperature intolerably high. Ironically, the total amount of heat dumped, per unit of electricity used, would be many times less than that resulting from the collection of solar energy on earth. Solar collectors are inefficient—10-15 percent goes into electricity, with all the rest heating the atmosphere—so if they are put into space, the waste heat will be harmlessly radiated away into space. Sunsats also beat nuclear and fossil-fuel plants in the game of waste heat reduction.

As we can see, there are a lot of red-herring issues raised concerning sunsat power. It is still possible that the microwave transmission of energy from sunsats in geosynchronous orbit may be judged environmentally unacceptable. It might be worthwhile then to look at alternatives: laser beams, space-manufactured hydrogen delivered to earth, or, if the materials technology could be developed, Arthur C. Clarke's "skyhooks," cables that connect geosynchronous satellites with the earth.

Nevertheless, the environmental battle over sunsat is on. The issue is joined, but the jury—hard data—is not yet in. It would be unfortunate if the debate were to become polarized immediately, before we are able to take a thorough, rational look at the data and compare the alternatives.

Meanwhile, we *can* say that if sunsats were to prove to be environmentally acceptable—as they appear to be—we would see, for the first time in human history, a clear economic path to living on the fertile stars. History makes it clear that economic incentives force bigger and better (and sometimes unfathomable) results. It is also clear that, sooner or later, we shall see the day when the fertile stars are settled. Sunsats could become the analogs to Columbus seeking new trade routes to the Far East, to Captain Bligh transplanting breadfruit from the South Sea islands to the West Indies, to the Pilgrims seeking new economic opportunity on the North American mainland. The settlement of space is as inevitable as the settlement of the Western world, and the best guess now is that sunsats will pave the way.

5

PART III

How Can We Do It?

Artist's concept of a shuttle-launch space station.

A mass-driver (left) can launch millions of tons of lunar materials for processing in space.

7. Return to the Moon

In December of 1972, two American astronauts blasted off the moon, leaving behind the remains of a brief, but unforgettable episode in man's history. The lunar modules, seismographs, roving vehicles, corner reflectors, and assorted other equipment now standing on the lunar surface will always be regarded—like the pyramids—as puzzling monuments, invulnerable to the passage of millions of years.

When will we go back to the moon? According to the current austere plans of NASA, not until after the year 2000—and perhaps never.

But recent developments suggest otherwise. In order to set about opening the door to the fertile stars, engineering studies point toward an early return to the moon as a surprisingly cheap and necessary step. These studies addressed the question of what is the smallest feasible facility that could transport, process, and manufacture useful products from lunar materials. The answer appears to be that a rapidly growing, self-replicating and cost-effective system could be built, launched, and landed on the moon with an investment of about $5 billion, about the level of NASA's current annual spending.

What needs to be lifted from the earth and built on the moon in order to begin launching thousands to millions of tons from its surface? The most likely candidate is a long, narrow electric motor that uses magnetic fields to accelerate bags of lunar materials in recyclable buckets to escape velocity from the moon. Energy to power the magnets would come from solar collectors along the track. At the end of the track, the lunar material would be hurled into space, collected at a convenient point, and taken to a factory in space. Each bucket would then return to pick up a new payload. A steady stream of bags, each containing about ten kilograms of moon dust, would amount to hundreds of thousands of tons of mass per year.

This device is called a *mass-driver.* One model recently built at MIT met specifications for use on the lunar surface; a second model being built at Princeton University will provide accelerations of up to 500 gravities. This means that a velocity of 250 miles per hour can be achieved in a distance of four feet!

After travelling for about one kilometer, the sacks would achieve escape velocity from the moon (more than two kilometers per second). Then, a few kilometers downrange, magnetic fields would guide the lunar material accurately along its path so that it doesn't miss its target 63,000 kilometers behind the moon. Engineering studies have shown that this fine tuning will enable a bullseye one meter across to be hit by each lunar bag.

After several thousands of tons are collected by the catcher over the course of weeks, the material will be nudged gently on a trajectory to a convenient orbit above the earth, about two-thirds of the distance to the moon. There it will be processed into satellite power stations and other products. The total energy cost of these transfers is very low, and the orbits selected have been found to be stable over long periods of time. There is no need to worry about their crashing to earth or escaping into the darkness of interplanetary space.

The workshop scientists and engineers found that the most attractive scenario involved first landing about 60 tons of equipment on the moon and another 90 tons of manufacturing apparatus into an orbit high above the earth. (In the next chapter, we'll see just what is required to get those 150 tons where they're needed.) The equipment on the lunar surface would include the components of a mass-driver, a processing plant that could convert lunar silicon into solar collectors in order to later expand the size of the mass-driver, some machines that could self-replicate while expanding the capacity of the mass-driver and processing plant, and a crew of three to oversee the growing operation. Raw materials launched from the moon would be processed into products in space—eventually into satellite power stations and space habitats.

With machines replicating themselves and their most massive products, rapid growth would take place. Just after the first landing, the study found, the initial lunar mass-driver could launch about 1,800 tons of lunar materials during its first year of operation. The doubling time for moving materials into space would be a mere 90 days, so that an astounding production rate of 100,000 tons per year could be reached after two years of bootstrapping. Also crucial to this scenario is the ability of the mass-driver and processing plants to produce about 100 times their own mass in one year. After three years, it would be possible to start building full-scale commercial sunsats.

The studies also found out that between 90 and 100 percent of satellite power stations and other heavy products could be fabricated from lunar materials. The leverage implied here is obvious: heavy machinery can be built almost entirely from lunar materials and the lunar and space enterprise can grow very fast, with relatively little replenishment from the deep gravity well of earth.

One can envision a lunar base with a rotating crew of three people monitoring the operation of the mass-driver, machines, processers, and a small bulldozer, which scoops up the lunar dust for export into space. There is no need to strip-mine the moon. The first million tons that could be launched over the first few years of the 1990s represent an excavation 30 feet deep times the area of a football field. It could not even be seen from a

(Above) At the lunar mining control center, a remote-controlled bulldozer is scooping up moon dust that will soon become fuel, solar collectors, and structures in space.
(Below) This astronaut is monitoring the trajectory of packets of lunar material being magnetically launched by a mass-driver.

telescope on earth, yet it contains enough material to build the first two or three full-sized satellite power stations, each capable of supplying a large city with all its electricity.

The first lunar inhabitants would live in cabins partly buried in lunar dust to protect them from the cosmic rays of outer space. Even when the satellite power program is in full swing, only a few people would need to be there at one time. The engineering studies have concluded that a total of only a thousand tons of equipment and personnel will need to go from earth to the lunar surface in order to launch hundreds of times that amount into space.

Even before the leap to satellite power and space settlements is made, a number of near-term benefits could come from lunar materials between 1985 and 1995. They could even pay off the investment *before* sunsats are built. In other words, sunsats alone may not be the necessary economical path to the fertile stars.

Among the possible benefits are the following: a several billion dollar savings in the delivery of oxygen for fuel for space vehicles and control systems for the support of crews; another several billion dollar savings in the delivery and construction of solar-cell arrays and radiators for science, weather, and communication satellites, solar electric systems to power these spacecraft, test modules, and satellite power demonstration plants; the availability of thousands to millions of tons of rocket fuel for orbital transfer of large cargos, for asteroid retrieval, and for other interplanetary missions; the availability of sufficient material for space-habitat shielding from lethal cosmic rays; a revival of lunar science and exploration; proof of technological readiness to transport, process, and fabricate nonterrestrial materials for space industrialization as an option to the earth-launch of all space facilities; and, finally, hard-to-anticipate or hard-to-quantify technological and scientific opportunities of possibly great potential, for example, silicon-crystal growth, vacuum deposition of metals, ore concentrations on the moon, and new discoveries in astronomy and planetary science.

What might previously have been thought of as an expensive and unnecessary revisit to a desolate world covered with gray dust could turn out to be the first giant leap to a civilization among the fertile stars.

8. Playing in Space

At this point, you might ask how we are going to place 60 tons of material on the moon and another 90 tons into a high earth orbit when one gigantic Saturn rocket could only raise one twentieth of that weight. How are we going to replenish water, food, people, and critical earth-manufactured components to meet the demands of a rapidly growing enterprise in space?

To get a feel for how much pushing is required to move the 150 tons where they need to go, imagine launching 20 adult elephants into a low "parking orbit" in space, followed by an additional spacecraft to boost all of them into a high orbit at the outer fringe of the earth's gravitational influence. To do this, the weight penalty for the new rocket and its fuel is three times what was needed to get the elephants into the parking orbit. Then, we wish to land eight of the elephants on the moon and eventually return some of them to earth. Yet another rocket and more fuel will be needed for this. The additional weight penalty is "two times," a factor of six over what was required to get the animals into space. Although the scale of these operations is elephantine, it is still thousands of times less than what would be needed for the earth-launch of sunsats.

The key to the elephant problem is the space shuttle, NASA's transportation system of the 1980s, which will enable us to begin to crawl out of our deep gravity well. As mentioned earlier, the shuttle is a manned rocket-plane the size of a DC-9, capable of launching 30 tons of cargo into orbit, returning to earth, and landing like an airplane. It is now being tested and is scheduled for its first orbital flight at about the time this book is published. Eventually, the United States is planning one flight a week for this big shuttle. I had been a critic of this project for a number of years, mainly because NASA hadn't come up with a goal well-matched to the tremendous capacity of the shuttle. Now there is a goal: launching the building blocks in space that will lead to the construction of sunsats and other products from lunar and asteroidal materials.

The space shuttle represents an increase in the total amount of material you can put into space by a factor many times greater than earlier launch systems. The engineering studies show that fewer than four shuttle flights per month for a few years, which is what NASA is planning, are adequate to begin and sustain the way toward the fertile stars. With each launch, we would fill the cargo bay with rocket systems, life-support systems, power plants, chemical-processing plants, lunar and asteroidal mining operations—enough to establish a nonterrestrial mining and processing capability. Then you could start building structures at a rate many times greater than that which originally came out of the deep gravity well of earth.

A chemical rocket—in NASA jargon called the *Interim Upper Stage*—is being planned for boosting payloads into high orbits. This rocket, plus a modified Apollo Lunar Module, could solve the "elephant problem" with a small development cost.

As an alternative, mass-drivers could also be used to boost payloads from low earth orbits to higher orbits and toward the moon and asteroids. As we

(Above) The space shuttle releases a satellite designed to look at the Earth's resources.
(Below) The shuttle glides to its landing strip at Cape Canaveral after a trip into space.

have seen, the shuttle can attain an orbit only a few hundred miles high, which is barely one-third of the way out of the earth's gravity well.

Prefabricated sections of the mass-driver and its solar collectors could be shuttle-launched from the earth and assembled in space. O'Neill has suggested that fuel for the mass-driver could be garbage-compacted debris, comprising spent shuttle fuel tanks. The mass-driver and its cargo would be tugged in an orbit spiralling outward in reaction to the expulsion of the unusual fuel. Eventually, large quantities of lunar and asteroidal materials could become available for mass-driver fuel. Using this bootstrap approach, we would be able to establish the first frontiers in space far away from the earth, getting the energy and materials for our space-rocket free (from the sun and spent fuel tanks).

During the buildup phase of the program, it would also be necessary to test long-duration human exposure to the space environment, concepts of chemical processing in space, and smaller satellite power concepts. Tinkertoy modules lofted aboard the shuttle could be put together in low orbits to construct testing facilities, crew quarters, and rocket systems.

The engineering studies envision a three-phase program. The first phase would comprise a U.S. federally funded effort over the next seven to eight years to demonstrate the feasibility of all the steps required to make space manufacturing a worthwhile option; most of this could be absorbed in programs already planned by NASA and the Department of Energy.

The second phase, during the late 1980s, would be the buildup period, during which 20 to 30 shuttle flights would deliver cargos into space. Then we would assemble the first lunar-mining station and the first full-scale chemical-processing plant. The economic incentives for making this happen could be nearer-term and smaller-scale than a sunsat program. One study by the Aerospace Corporation concluded that there will be a need for significant expansion of earth resource and communications satellites that would weigh up to tens of thousands of tons.

Private and international investment would begin to bear the brunt of expenditures during this period, when several tens of billions of dollars could be parlayed into an even larger marketplace by the turn of the century.

The third phase would be the commercial use of satellite power, which could begin as early as the 1990s, when tens of billions invested could also lead to hundreds of billions returned.

During the buildup period, the crew quarters could be quite attractive. An individual office/bedroom would be one of the 20 to 30 sizable pie-shaped slivers tucked inside a spent shuttle fuel tank. The tanks could be

(Above) Astronauts are putting together shuttle-launched pieces of a mass-driver rocket that will lift cargoes from low orbits into high orbits and to the moon and asteroids.
(Below) The crew quarters can be pleasant during the early stages of space settlement.

tethered to one another and spun up to create artificial gravity. Toward the center, there would be space for low- and no-gravity recreation. Living and working in space can be a pleasant experience, even during the pre-colony stage envisioned for a decade from now. Fairly early on, we could abandon the battleship-gray, antiseptic environments of *Star Wars* and *Star Trek*, and we could begin to take on qualified professional people outside the confines of the military and astronaut images. *All this could happen within the next ten years if we wished to do it.*

9. Space Odyssey to an Asteroid

This is a scenario for asteroid retrieval: Space Odyssey 1997, occurring 505 years after Columbus's historic voyage. For the first time in history, humans are exploring new worlds beyond the moon, which had been visited only 28 years earlier in the Apollo program.

This time the astronauts are embarking on a much bolder venture. They are preparing for a landing on an asteroid called Columbus, at the time situated 30 million miles and six months' journey from the earth. A crew of 12 will set up a mining operation on the asteroid, an irregular, potato-shaped rock looking like Viking photographs of Phobos, a moon of Mars. Columbus—called *1988 BD* when it was discovered nine years ago—is only about 100 meters across and weighs two million tons, barely the size of the smallest crater discernable on a photograph of Phobos. The crew is giving the mass-driver a final check before spiralling out of the top of the earth's gravity well toward the asteroid. The mass-driver is using lunar material for fuel.

When the astronauts arrive at the asteroid, they attach small rockets and cables to stop it from spinning. Then they hollow out a cave inside the asteroid for their living quarters. It will protect them from constant exposure to cosmic rays that could cause cancer.

The crew then puts a bag around the asteroid and attaches it to the mass-drivers, which propel asteroid particles to very high speeds, nudging what remains of Columbus gently toward a new path around the sun. After six months, they fly by the planet Venus, which will give them a gravity assist. The astronauts will now ride the asteroid toward the earth. Solar collectors along the noodlelike tubes of the rockets—which are several miles long—are continuously soaking up sunlight to produce magnetic fields that accelerate the particles inside the tubes. The free energy from

An Earth-bound asteroid has been captured, bagged, and is being towed by three mass-drivers past the planet Venus.

Two tractors are chewing up this asteroid for fuel while the astronauts live inside hollowed-out caves.

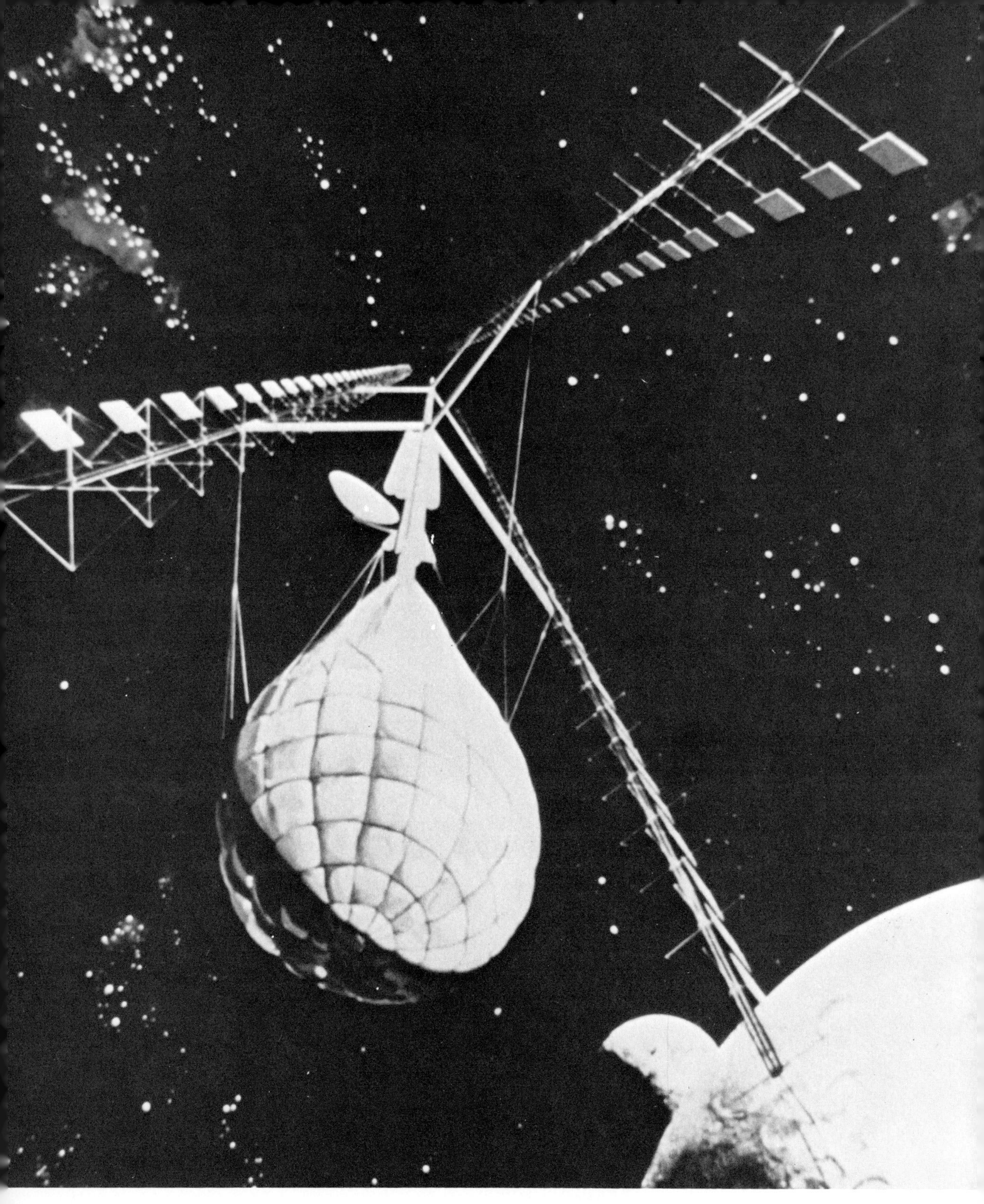

After its long journey, the asteroid passes the Moon and is captured into an orbit circling the Earth, where it will be processed into large space structures.

the sun and matter from the asteroid are the only resources needed to move the asteroid to a high orbit around the earth.

The greatest challenge in this mission has been to send the crew and retriever to rendezvous with the asteroid. The buildup was done over one or two years of space-shuttle launches to earth orbit, where the mass-driver, mining apparatus, and crew quarters were assembled. The asteroid-retriever then made a slow spiral out of the earth's gravity field on a half-circle around the sun to the rendezvous point.

Small mining tractors are now chewing up Columbus at the rate of eight kilograms per second. The loose agglomeration of material is rich in carbon and water, like the carbonaceous meteorites which strike the earth. A large mirror concentrates sunlight to heat up the asteroid dust. Water and carbon are driven out by the solar furnace and stored as ice and dry ice in bags on the shady side of Columbus. Iron and nickel particles are extracted by magnets, and the rest is used as fuel for the mass-driver.

Near the end of the mission, the asteroid passes the moon, which assists it in its final capture by the earth. In three years, about 400,000 tons of ice, dry ice, and metals will begin to be used by hundreds, and later thousands, of people inhabiting a modular space station orbiting the earth at two-thirds of the distance to the moon. The remainder of the asteroid, another million tons, will be used for further processing, for habitat cosmic-ray shielding, and as reaction mass (fuel) for later journeys to larger asteroids.

These new space settlers in high earth orbit, while awaiting the arrival of the asteroid, are preparing to construct enormous solar collectors and microwave transmitters to supply energy to the earth. The collectors will be made of asteroidal materials: metals and glass processed in space using continuous solar energy. There will be sufficient material from this first asteroid, about one million tons, to build collectors for supplying the United States with much of its increased electricity demand around the year 2000.

Toward the turn of the century, larger asteroids will be retrieved, and nuclear, coal, and oil plants can then be phased out as the primary sources of central-station, base-load (continuous) electricity on earth. Dwindling supplies of nickel, iron, and other metals on earth will be replaced by asteroidal metals. Huge agricultural areas in space, built from asteroidal materials, will grow food that will be shipped to earth to relieve the starvation or undernourishment of billions of people.

Returning to our scenario, after four years of nudging by mass-drivers and by the gravities of earth, Venus, and the moon, the asteroid Columbus is in its final orbit, being processed into satellite power stations. The chemical-processing plant here, which weighs about 2,000 tons, was either launched from the earth by space-shuttle flights over a year or two and boosted to high orbit by mass-drivers that chewed up spent shuttle tankage as fuel, or, if a visit to the moon has been made, it was made of lunar materials. This factory is producing hundreds of thousands of tons of products from the asteroid, so there is a hundredfold return on investment from a few thousand tons of mass-drivers, mining machinery, processing plants, and crew quarters.

Industry has moved into space, and it will be only a matter of time before the retrieval of much larger asteroids will be possible. The one-hundred-times leverage is just the beginning. After Columbus has been returned, the door is open to a virtually infinite storehouse of fertile planetoids.

Science fiction? No, not according to the sober engineering studies that have been made of this scenario. It will be challenging but it is not beyond the realm of currently available technology, and it would not break the bank. Asteroid-retrieval missions on the scale described here would require smaller investments than the Apollo program. If, for some reason, the mass-driver does not prove to be reliable or economical, there are several alternatives: chemical rockets, using asteroidal hydrogen and oxygen as fuel; ion engines, using asteroidal silicon or oxygen as fuel; or solar sails, huge, but easy-to-build, thin, lightweight aluminized sheets that can travel with their payload around the solar system using the pressure of sunlight as fuel. And we can't forget redundancy: backup, or separate, systems that vastly increase the probability of success. Columbus need not be a one-shot affair.

A number of ingenious ideas have been proposed to minimize the cost of asteroid retrieval. Dr. Cliff Singer of Princeton, for example, has suggested that a column of debris from one asteroid could be diverted toward a second asteroid so that the second asteroid's orbit could approach the earth's (or Venus's), whereupon gravity assists plus a little more fuel could take the raw materials where they are needed. Calculations have shown that asteroid pairs could be found where the total amount of fuel, and therefore the cost, would be surprisingly low.

Another idea, explored by Dr. David Ross of Stanford University and the Jet Propulsion Laboratory, is to "put the brakes" on incoming asteroidal material by means of a double lunar capture, where the material swings by the moon, then by the earth, then back by the moon, whereupon it becomes captured in the earth's orbit.

Scott Dunbar at Princeton is searching for asteroids located near a point forming an equilateral triangle with the earth and the sun: in other words, in the same orbit as the earth, but 60° apart from both the earth and the sun. There are a number of reasons we suspect they might be there. The

French mathematician LaGrange predicted "stabilities" for asteroids venturing into these special regions with respect to a planet and the sun. In fact, Jupiter and the sun possess a number of trapped asteroids called *Trojans*. If earth-sun Trojans also exist, the energy required to retrieve them would be tiny.

But before we can undertake a retrieval mission to an asteroid, we would like to know more about their chemistry. We suspect we will find many attractive candidates, both from the point of view of accessibility and materials. The best route is to significantly expand the asteroid search program (an inexpensive thing to do) and to begin a program of unmanned missions to rendezvous and sample prime candidates (still at a modest cost for NASA—one shuttle flight containing an upper stage rocket could go to the asteroid 1943 Anteros in 1992 and return a one-kilogram sample to Earth). These were the recommendations to NASA made by two scientific study groups during the summer of 1977.

10. Comets as Fertile Stars

Comets are tiny, icy worlds coming from the dark outer reaches of the solar system. Astronomers believe there are billions of them inhabiting a vast invisible reservoir, orbiting the sun out to a distance of one light year. They theorize that every few million years a star passes close enough to perturb a small fraction of them into highly elongated sunward trajectories.

When a comet travels inside Jupiter's orbit, the ice begins to evaporate from its surface to form a large cloud of gas and dust around the small, solid nucleus. Sunlight and solar particles sweep some of the gas and dust away from the comet and the sun, forming a tail up to millions of kilometers long.

Jupiter and other planets can capture some comets into orbits that bring them close to the sun every few years or decades. These comets are doomed to disintegrate completely after one or more solar passes. But most comets that come near the sun disappear into the frigid interstellar reservoir from whence they came, never to return again.

Comets are probably the most primitive objects known to visit the earth's neighborhood. The residue of a condensing solar nebula, they appear to be unscathed by the drastic changes that gravity and sunlight have inflicted on planets, satellites, and asteroids. The rocky and icy materials in comets may also contain significant amounts of organic materials, including amino acids. Some scientists suggest that collisions of comets with the earth have provided our planet with organic precursors to life.

Comets can create a dazzling display in the sky; they caused some terrifying, myth-producing spectacles during ancient times. Comets are almost certainly the parent bodies of many meteors that streak into the earth's atmosphere and disintegrate. The most famous historical example of what scientists believe was a comet-turned-meteor was the devastating 1908 Tunguska event in Siberia, where a 60-meter-diameter, icy chunk exploded above the pine forests, releasing energy equal to a hydrogen bomb.

In the future, comets will be the destinations of space missions, first for scientific reasons, later as convenient stepping-stones on voyages to the stars.

The most famous comet of all is Halley's Comet. Returning near the earth and sun every 76 years, it is due for another visit in early 1986. At present, Halley's Comet is drifting slowly sunward from Uranus's orbit. It is on the threshold of being observed by the most powerful telescopes. Although astronomers have searched for it, they have not found it yet, suggesting that the solid part of the comet is no larger than 7-12 kilometers in diameter.

It will be another five years before Halley's Comet goes on stage for its once-in-a-human-lifetime appearance. Scientists are now preparing an unprecedented assault on Halley's Comet; it will begin with the first glimpse of the faint, inert dot that will soon appear in somebody's telescope. Halley's Comet is the only large comet predicted to return this century, so the effort is well justified.

The International Halley Watch, to be headquartered at the Jet Propulsion Laboratory in Pasadena, California, will coordinate space and ground observations of the arrival of the great comet. Delays in the space shuttle and tight budgets are making American plans on-again, off-again, but the Europeans are prepared to go it alone if NASA backs out.

If all goes well an American comet probe will be launched aboard the space shuttle in July, 1985. In November, it will fly by the comet at the staggering speed of 60 kilometers per second, still slow enough, however, for its cameras to glimpse the comet's tiny nucleus. The spacecraft will also deploy an atmospheric probe that will streak through the cloud of gas and dust surrounding the nucleus, scooping up what might be our first pristine sample of the interstellar soup from which the solar system was formed.

The spacecraft will then loop around the sun out to the asteroid belt and come back in to rendezvous with the comet Tempel 2 in 1988. Scientists

are planning a detailed one-year observation of Tempel 2. The Europeans, by contrast, plan only a 1985 Halley flyby—but at least they have a commitment.

With the November 1980 Voyager encounter with Saturn we are entering a hiatus of at least several years in solar-system exploration. The proposed comet mission may wake us up from our long sleep; the 1980s could well be remembered as the decade of the comet. In addition to the possibility of NASA participating and the certainty of the Europeans launching a Halley probe, the Soviets and Japanese are considering comet missions, too. It is hoped that the results can be coordinated to develop a complete, coherent scientific picture of comets.

Engineers are as excited as scientists about a mission to Halley's Comet. The long voyage would be the first practical application of NASA's solar-electric propulsion, which would gently thrust the spacecraft past Halley's Comet on to the rendezvous with Tempel 2. Large solar arrays would provide up to 3 kilowatts of power to a high-velocity mercury-ion thruster.

Imagine yourself as an ageless inhabitant of Halley's Comet, able to take a good look at what earthlings are doing during each 76-year visit. Chances are you would not have noticed the explosion of technology until your 1910 visit, when you would have observed small airplanes, cars, electricity, telephones, and some fair-sized telescopes looking up at you.

Your 1986 appointment would shock you: nuclear bombs, jumbo jets, freeways, skyscrapers, and space ships, one of which streaks through your atmosphere at kilometers per second.

What will happen when you come back in 2061? It is not far-fetched to visualize being colonized by a group of adventuresome inhabitants of earth, since comets are made of materials essential to life and they are everywhere. The visionary physicist Freeman Dyson points out that the combined surface area of all the comets is at least a thousand times that of earth and suggests that they, not planets, may be the major potential habitats of life in the solar system. They inhabit the vast reaches of interstellar space like oases, conveniently separated by light-days, not light-years, and they will surely be visited by the first travelers to the stars.

11. A Unified Scenario

You might be confused at this point, lost in a maze of scenarios, timetables, destinations, tonnages, costs, rocket varieties, factory types and so forth. Engineers, scientists, and economists seem to agree on the feasibility and cost-effectiveness—and fun—of lifting ourselves from the bondage of the earth's gravity and of processing fertile stars into usable products that could alleviate the stresses we feel here on earth. But a consensus on a particular plan of action has not yet been achieved, and there is a lot of misunderstanding as a result. Furthermore, NASA has not yet seen fit to look at the entire array of low-cost precursor activities.

In a rational world, we expect agreement on the best-known scenario of the moment, subject to revision as new pathways and goals are discovered. But since the world is not rational, we need to go along with what we have: a blend of individuals with their own preferences and prejudices; a changing political and social climate; and technological innovation to surprise even the most educated minds. We are left with guesswork, guided by a blend of logic and hunches and a little extrapolation into the future.

I would like to offer my own 1980 view of the most likely pathway to the fertile stars. My scenario unites a number of apparently conflicting assumptions: a logical and low-cost sequence of events; the inevitability of the events; NASA's current austerity; no significant technological advances; the construction of satellite solar-power stations, space settlements, and other products composed of nonterrestrial materials; and the acquisition of food and raw materials from space for consumption on earth.

The scenario, developed after a decade of work in the field and from interviews and experiences, is outlined below:

1980-85: Increased public education and awareness of the severity of the energy, food, and resource problems; a change in political leadership; the first operational successes of the space shuttle; and a few key low-cost feasibility experiments on earth and in space (the cost above NASA's current estimate: about $1 billion).

1985-90: The deployment of 150 tons of cargo and small crews to high earth orbit and the moon; and the creation of the first lunar materials-processing facility, with an immediate return on investment in the production of oxygen for rocket fuel, silicon for solar panels, self-replicating machines, larger resource and communication satellites, etc. (the cost above NASA's current estimate: about $5 billion).

1990-95: Using lunar bootstrapping, begin the retrieval of millions of tons of material from an earth-approaching asteroid or Phobos; construction of the first commercial sunsats; and the first space habitats and agricultural areas (an investment smaller than or equal to the Apollo program, depending on the pace of development).

1995-2000: Bonanza! The lifting of the limits to growth on earth; and the beginnings of human civilization among the fertile stars.

PART IV

What's In It For Us?

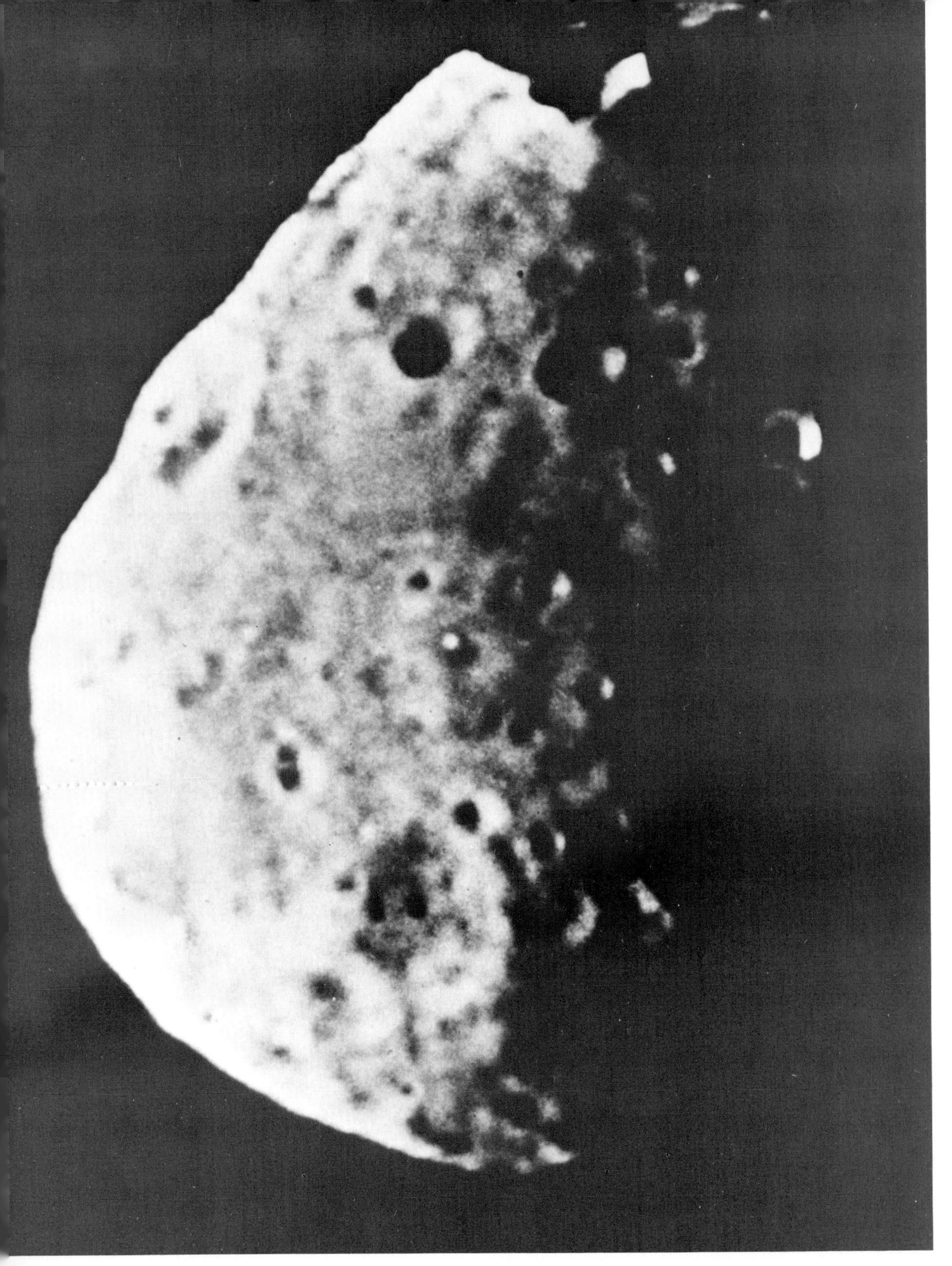

Phobos, the rocky moon of Mars

12. *Watch Out for Falling Fertile Stars!*

On June 30, 1908, a 10-megaton "bomb" exploded 5 kilometers above the pine forests of Tunguska, Siberia. Trees fell and branches sizzled over an area 60 kilometers wide. About 1,500 reindeer were killed and a man standing on his porch was knocked down. Shock waves registered on seismographs and air-compression waves were detected in Europe, thousands of kilometers away. A brilliant fireball was observed in broad daylight.

The explosion released as much energy as a sizable hydrogen bomb. What was the cause? Either a small comet or an extinct comet-turned-asteroid was the culprit. The loose agglomeration of dust and ice, about 60 meters across and weighing 100,000 tons, exploded in the air, leaving a horrifying disaster in its wake. This natural event, whose potential for detruction far exceeded a thousand falling Skylabs, fortunately took place in an uninhabited area. When are where this sort of thing will happen again is open to the speculation of disaster scenario writing (for instance, the recent movie *Meteor* and the often-repeated TV movie about a comet that exploded over Phoenix). We can be certain of one thing: events of this magnitude or greater do occasionally happen, and cataclysmic collissions with larger comets, asteroids, and meteorites may even be a major factor in changing the climate, geology, and evolution of living things on earth.

Meteor Crater, near Winslow, Arizona, is an example of the impact of a small asteroid in geologically recent times, about 22,000 years ago. The crater resulted from the impact of an iron-rich asteroid, 50 to 100 meters across and weighing several hundred thousand tons (slightly smaller than the hypothetical asteroid Columbus described earlier).

How abundant are the earth-approaching asteroids and how likely are we to get hit by one? Estimates of the numbers and sizes of craters on the moon and Mars, combined with data on the Apollos and Amors whose orbits are known, suggest that there are at least 100,000 earth-approaching asteroids with diameters greater than 100 meters and weighing a million tons. This is approximately the size we assumed in our retrieval scenario, and the size of the body that produced Meteor Crater in Arizona. Many of these asteroids are probably carbonaceous; some may be extinct comets (objects that have come in from the outer reaches of the solar system and have been perturbed into earthlike orbits); while others are metal-rich, like some of the meteorites that have struck earth.

Scientists who study changes in asteroidal orbits around the sun have concluded that all currently observed Apollo and Amor asteroids will eventually impact earth, Venus, or the moon, and most of the events will occur over the next 100 million years. This is attributed to the unrelenting and uneven pull of the planets.

Do we need to run for cover? Given that small meteors are more common than larger ones (few people realize that more than a ton of unseen meteorite dust falls on earth every day), how big does a meteor have to be to cause havoc like the Tunguska event, and how often does this happen? What would be the loss of life and property?

These were questions posed by a group of scientists recently convened by Dr. Bruce Murray, director of the Jet Propulsion Laboratory in Pasadena, California. The conclusion was that more information needs to be gathered to assess the hazard, but there exists an uncertain, low probability that a very destructive event could occur in our lifetime, where the loss would include many lives and billions of dollars of damage.

Princeton physicist Theodore Taylor speculated on the losses that would result from the impact of an asteroid of 1 kilometer diameter, an event that befalls the earth every few thousand years or so. Although the probability of such a catastrophe in our lifetime is very low, the devastation would be far greater than we might imagine: enormous tidal waves, atmospheric changes that could create climatic havoc, and the levelling and burning of thousands upon thousands of square kilometers. The loss could run into the trillions of dollars; in fact, civilization as we know it might well cease to exist. A war of hydrogen bombs would be tame by comparison.

Applying logic similar to that used in studies of nuclear power plant safety, where the probability of a catastrophic accident is very low but the consequences very severe, Taylor suggested that a billion dollars of prevention may be worth a trillion dollars of cure. (Regardless of which side of the nuclear-power issue you are on, everyone agrees it is worthwhile to make existing plants as safe as possible—and that costs money.)

What can we do to prevent an asteroid apocalypse? The best answer is to be prepared to divert the asteroid. Reminiscent of the movie *Meteor* (although we are now talking about serious, precautionary engineering plans), we would need to develop a system that could be deployed quickly to alter the path of the asteroid away from earth. There would normally be at least a few days' notice, perhaps more if the orbit of the asteroid were already known.

What type of asteroid-diversion system is needed? The same kind we need to utilize asteroidal resources. *Another* rationale has been added to opening up the fertile stars: that "ounce of prevention" might provide the means to develop the technology for retrieving asteroids. Sometimes a good thing (improvement of the human condition) is done for a questionable reason (here, exaggerated fear). But in our irrational world, the repetition of a Tunguska-like event might be what it will take to precipitate a clarion call to action in the congresses and councils of the world's governments.

Whether or not we go to the fertile stars, they will come to us.

13. Mother Lode from Space

About two billion years ago, probably before life appeared on earth, an asteroid crashed into the plains of Ontario, not far from the mining town of Sudbury. About one-half of the world's nickel comes from the large, circular crater formed by the impact, but geologists disagree about whether this rich cache of nickel (and iron) is literally extraterrestrial, or whether the impact excavated subsurface ore. We may have been unwittingly mining a fertile star!

Mining the asteroids for their metal resources is not a new idea, but that it could be done economically using current technology *is*. We have considered the standard scenario of going to the moon, using lunar materials for fuel, and constructing space vehicles, power sources, and satellites, leading to a cost-effective program of satellite power. There is another scenario that could occur, either independent of or as a consequence of the satellite-power scenario: sending selected asteroidal (and perhaps lunar) metals to the earth's surface at competitive market prices.

Astronomers Michael Gaffey and Thomas McCord have suggested that an iron-nickel asteroid could be cost-effectively recovered on earth to replenish the world supply of these metals. The material could be benignly recovered in the ocean and towed ashore by barge after reentry into the earth's atmosphere in metal-foam bodies made in the vacuum of space.

Some meteorites that land on earth are almost entirely iron, nickel, or other metals. Spectroscopic studies show that some of the asteroids are likely to be similar to these museum specimens—chunks of ore made mostly of iron, with about 10 percent nickel and 1 percent cobalt. The prices of nickel and cobalt are several dollars per kilogram, with a world market of billions of dollars per year. So if the price of asteroid retrieval could be lowered to a dollar or less per kilogram, the extraction of asteroidal cobalt, iron, and nickel could provide an economic justification for venturing to the fertile stars.

Both the mass-driver scenario and the solar-sail scenario of MIT scientist Eric Drexler for retrieving large tonnages of asteroidal metals seem to be feasible. Drexler pointed out that asteroid-mining might be justified by using asteroidal steel in space in radiators to dissipate waste heat from satellite power systems and as shields or hardeners for military satellites.

Hundreds of thousands of tons of metals may be required for these systems during the 1990s, providing yet another practical incentive for asteroid-mining.

Less abundant precious metals may also justify mining the moon and asteroids. Mining engineer David Kuck has investigated the amounts of rare metals in lunar and meteorite samples and concluded that titanium from the moon and platinum from some meteorites could be extracted in significant quantities, worth billions of dollars on earth. A 100-meter-diameter asteroid like Columbus would be big enough to do the job. Kuck's assessment, like Gaffey's and McCord's, is that the benefits would far outweigh the cost.

The science-fiction writer's dream of a gold rush in space may soon become reality. The metals in asteroids could provide the impetus for sober-minded mining engineers to join NASA and launch an extraterrestrial mining program, even if, for some reason, the sunsat concept were to fizzle out.

14. Breadbaskets in the Sky

The ability of the world to provide enough food and other resources for its growing population is becoming an increasingly serious problem. There appears to be a consensus among most investigators that widespread famine, ecological stress, and political and economic turmoil are the inevitable consequences. The Food and Agriculture Organization of the United Nations has often reiterated its warning about the long-term food crisis: "The basic world food situation remains as insecure as before, with the long-run production trends still inadequate."

The problem is complex, and lasting solutions are elusive. In a recent article, Lester Brown stated the problem in these terms: "The tripling of world economic output since mid-century has raised pressures on biological systems, often to an unsustainable level. . . . In economic terms, we are now consuming biological capital."

The evidence Brown cited for this decline included the 1972 Soviet wheat purchase and the subsequent leveling off of world grain production, rising energy costs increasing reliance on marginal lands, a deteriorating crop base, the decreasing effectiveness of fertilizers, the decreasing availability of water for irrigation, the increasing demands for land for nonagricultural purposes, soil deterioration, the leveling off of fish production due to overfishing, overgrazing of grasslands, and the decimation of forests—all in the face of a growing world population.

The most relevant question in addressing the world food problem, according to agronomist Roger Revelle, is this: "Can rates of growth of agricultural production be made to exceed rates of population growth, thereby improving the condition of life of poor people throughout the world? Such improvement is probably one of the essential conditions for reducing birthrates and eventually stopping population growth."

The challenge of increasing the world's food supply in lasting and environmentally compatible ways involves increasing both the resource base and the efficiency of distribution. With regard to the former, Revelle cited several fundamental limitations and causes for concern: the maximum sustainable fish catch is probably not more than twice the present catch; ocean farming is "not likely to add substantially to the food supply in the foreseeable future"; the increased use of fossil fuels may produce effects on the world's climate that cause serious disruptions in the food system; water resources are becoming more scarce; the growing demand for fossil fuels for agriculture is increasing its capital-intensiveness, making it more difficult for developing countries to keep pace; and, of course, the worldwide supply of fossil fuels will eventually run out.

Unfortunately, the development of modern methods of intensive agriculture—the "green revolution"—has not produced lasting solutions, although in some instances they have bought time. "The green revolution," wrote Jennings, "did not solve the problems of world food supply; rather it demonstrated an approach to a solution, a method."

Barry Newman reported in the *Wall Street Journal* that initially increased rice yields in Java later were reversed. "There isn't anything left in the Green Revolution's bag of tricks," he wrote. "The revolution, in fact, has turned against itself. Now that Java's paddies are supporting two or three crops a year, rice-ravaging pests can feast full time. The only deterrents are insect-resistant seeds, but the insects seem to get used to these seeds as fast as the scientists can fashion them. The new seed varieties are also more vulnerable to the droughts and floods that intensify on Java as erosion gets worse." Rice production in Indonesia has also dropped in the past few years. Referring to the situation in his native India, Salam wrote, "The chemical revolution of fertilizers and pesticides in agriculture touched us not."

Matters could get worse. "Expanding human population and changing agricultural technology threaten the genetic reservoirs of the world's major food crops," reported Garrison Wilkes in *The Bulletin of the Atomic Scientists.* The resulting paradox in social and economic development is that "the product of technology (plant breeding for yield and uniformity) displaces the resource upon which the technology is based (genetic diversity). . . . This in itself is not bad; but it does create a liability because the environment never remains the same." The margin of error in food production has decreased qualitatively as well as quantitatively.

This pessimistic view is shared by the Cornell agriculture scientist Neal Jensen, who investigated wheat yields in developed countries. "The dramatic increases in wheat yields that began in the 1930s in the United States will soon begin to level off," he wrote. "The favorable mix of genetics and technology that has characterized this era must build on an ever higher yield base for the future. At the same time, the residue of factors that can lower wheat yields includes a larger proportion of forces not easily shaped or controlled by man. An example is weather. The result is a natural yield ceiling that is already visible and that will impose a limit on future productivity growth."

It appears that the time may have come when the dire predictions made in 1798 by Thomas Malthus could become reality. Malthus argued that the growth of human population would proceed geometrically while food production would increase arithmetically, ultimately reaching a level of per capita production that would be inadequate.

Curbing the growth in population is not easy to achieve or maintain: "It is generally conceded," said Jensen, "that world population cannot stabilize below 12 billion people and may go as high as 25 billion people. . . . A favorable or desired trend in population stabilization must be sustained for something like 70 years for the entire population to reach equilibrium throughout its age structure."

Nevertheless, it has been suggested that a population of some 40 to 50 billion people, or ten times the current world population, could be fed, in principle, using the methods of intensive agriculture and converting previously unused land for growing food. Also possible is the widespread use of controlled environmental agriculture (CEA), where food is grown under carefully controlled closed or semiclosed conditions. But all these developments are capital-intensive and appear to be unattainable in the foreseeable future because of the complexity and shortsightedness of economic systems.

The problem of more immediate concern is the difficulty of achieving an equitable distribution of food products in a hungry world. The marketing system is extremely complex, and vested interests in profit are powerful. The unequal availability of fertilizers, fuel, and water and the remoteness of some potential customers exacerbate the problem. The problem for the United States in trying to feed the world, admitted Tom Wyman, president of Green Giant, is that getting the food from here to places like India is out of proportion to the payoff.

Bernard Feld put it more bluntly: "American food policy is rapidly becoming a scandal—and a tragedy. As the world's foremost food producer, we should be taking the lead in building up a stockpile of food as insurance against the inevitable shortages that accompany droughts, floods, poor planning, and other natural calamities. Instead, the only national food policy we have is to encourage profitable sales of excess supplies to the highest (or sometimes, the quickest) bidder."

The solution to the world food problem requires a radical shift in our ability to increase production and to provide a means for cost-effective, widespread distribution. The techniques employed in controlled environmental agriculture, applied throughout the world, may be the only long-term recourse, yet the rules of politics and the marketplace may prevent widespread implementation on any reasonable time scale. A food crisis may be looming with a potential for human hardship that makes the energy crisis appear to be mild in comparison.

"I suggest to those whose business it is to make projections on the world stage that absolute limitations to food production loom in the future," Jensen concluded. "We have been surprised at the rapidity with which the energy crisis, the depletion of fossil fuel supplies, came upon us. It would be tragic indeed for this to be repeated with food. The bicentennial of Malthus's paper will be in 1998. Let us hope that by that date the problem, if not the solution, will be much clearer."

The consensus is overwhelming: the world food problem is of immense proportion and does not appear to be solvable with the approaches we can envision. In late 1979, The Presidential Commission on World Hunger concluded that food, not energy, is the world's top problem. "The world hunger problem is getting worse rather than better," they wrote. "A major crisis of global food supply—of even more serious dimensions than the present energy crisis—appears likely within the next 20 years, unless steps are taken now. . . . Moral obligation alone would justify giving highest priority to the task of overcoming hunger."

The finite planet earth is truly on the verge of a breakdown, and food appears to be the precipitating factor. Farming the oceans, tinkering with improvements in distribution systems, and intensifying agricultural production are simply not enough. We will inevitably look away from our planet for the answer.

What are the prospects for supplying a hungry world with enough food in the year 2000 and beyond? It may come as a surprise that food grown in space and recovered on earth could provide the elusive solution to the food crisis. The means for expanding food production significantly could be well underway by the time of the bicentennial of Malthus's prophetic statements.

Is growing food in space feasible? Yes, according to experiments in the Soviet Union. The Russians have isolated people for up to six months in

closed environments, where they have successfully grown wheat and made bread. And they have done some experiments in space as well, to prepare them for long-term, agriculturally self-supporting, orbiting settlements.

A number of scientists in the United States have investigated the feasibility of carrying on intensive agriculture in space to supply food for the inhabitants of space settlements. Grains, bread, poultry, and pigs could be raised in closed agricultural areas adjacent to orbiting colonies, where light, temperature, and moisture could be varied according to the requirements of a particular crop.

Space inhabitants could develop the full complement of crops and livestock instead of resorting to the dullness of dehydrated foods and Tang, about which the astronauts have complained. Drought, pests, and pervasive disease—even the passage of seasons—could be eliminated. Fertilizer could be produced in space by using solar heat to combine nitrogen and oxygen from the asteroids. Water bound in asteroid minerals could be used as the major constituent for growing plants, perhaps by hydroponics, and

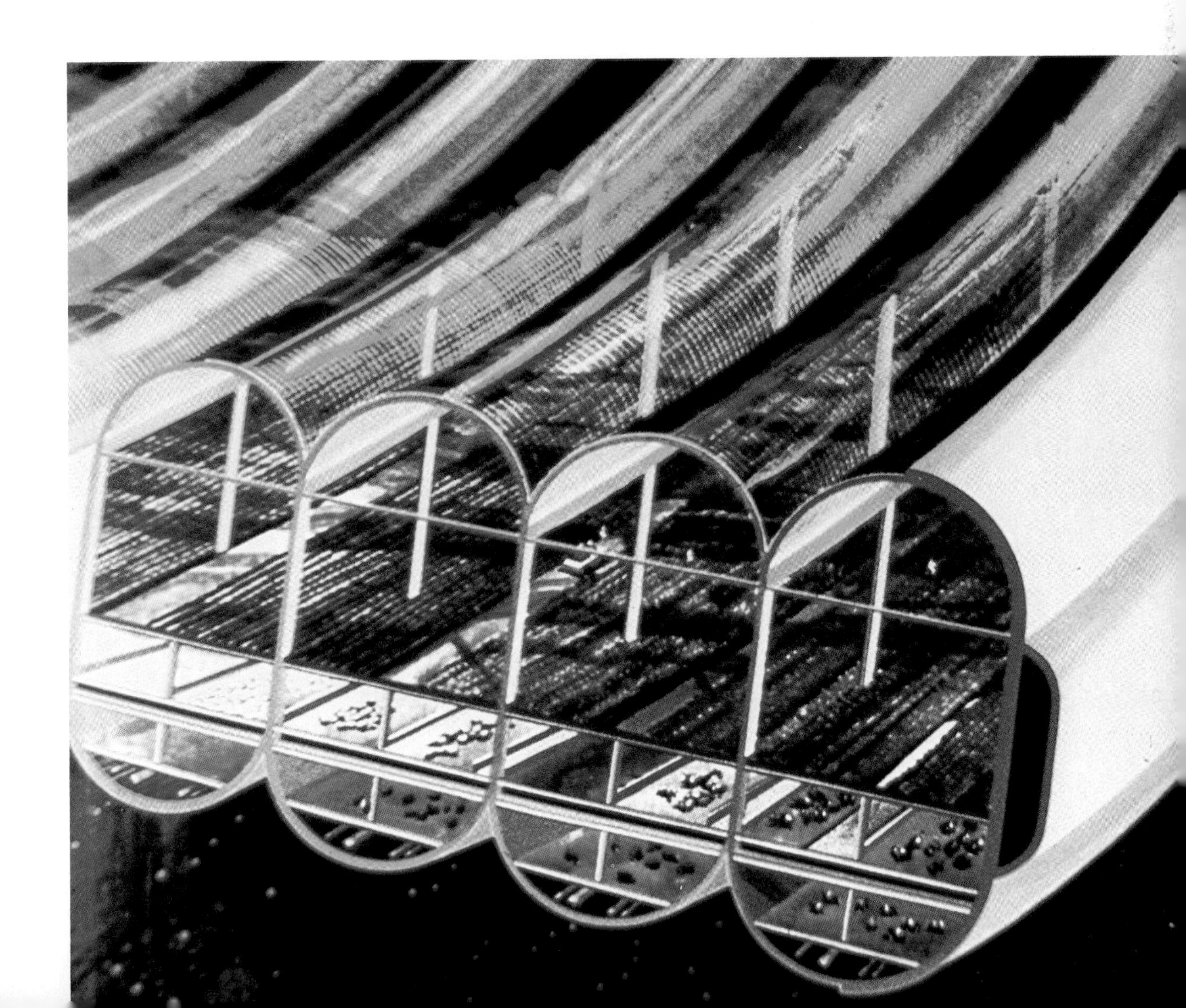

for sustaining humans and animals. (The cost of existing launch systems would make shipping the water up from the earth prohibitively expensive.)

The supply of materials available for agricultural facilities in space would increase rapidly as the number of people in space settlements grows, and eventually it will be possible to construct huge areas for growing food in space. Although the workability of closed agricultural ecologies has not yet been conclusively verified, it appears likely that, sooner or later, large amounts of food will be produced in space.

Recent engineering studies have shown that by the turn of the century our most cost-effective and environmentally compatible means of supplying electricity may be to build satellite solar-power stations and send their output to earth via microwave links. The stations would be built in space from materials retrieved from the shallow gravity wells of the moon and asteroids. Millions of tons of glass and metals would be required to build the power stations.

The number of people needed in space to support such an operation is debatable, but it is likely that there will be a good many of them. The 1977 Ames Summer Study projected that 3,100 people could be in high orbit by 1991 to process nonterrestrial materials into satellite power stations. The study concluded that about 30,000 tons of water, carbon, and nitrogen would be required to support this settlement and to establish food-growing areas. It would be extraordinarily expensive to launch these materials from the earth, and they appear to be almost totally lacking on the moon, so we are led inevitably to asteroid agriculture.

Science fiction? Again, no. We have seen that the technology exists to economically nudge asteroidal chunks into high orbits for processing into space settlements, factories, satellite power stations, and food growing areas. We have also seen that asteroids contain the vast bulk of the mass required to do the job. The economic incentive for producing satellite power may be sufficiently compelling to make space settlements and large-scale space agriculture a necessary outcome.

But even if satellite power, for some reason, does not turn out to be the primary incentive for developing the resources of space, some scientists and engineers believe that a lunar mining program could be economically justified to support the relatively modest projects planned by NASA for the 1980s and 1990s. For an investment of $5 billion to $10 billion, we could emplace a lunar mining and launching facility and a space chemical processing plant capable of producing hundreds of thousands of tons of lunar products: oxygen for rocket fuel and life-support systems; silicon for solar collectors; and metals for the supports and hulls of large space structures.

We are led to this logical conclusion: the door to large-scale manufacturing and agronomy in space is likely to open sometime during the 1980s, after the space shuttle starts routine flights and the Soviet Union establishes its first permanent orbital space station.

But what does this have to do with supplying food to earth? The favorable economics of asteroid retrieval, which is in the range of tens of cents investment per kilogram brought back and processed, combined with the inevitability of space agronomy, raise the possibility that food could be grown more cheaply and reliably in space than on earth. Large quantities of dehydrated crops could then be dropped out of orbit (possibly by an electromagnetic mass-driver device), enter the earth's atmosphere aboard a metal-foam reentry body, land in the ocean near potential consumers, and be towed ashore for use.

Science fiction? Again, no. It costs very little to bring things down from

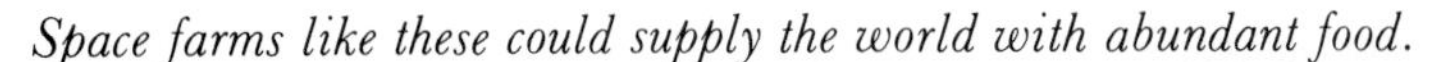
Space farms like these could supply the world with abundant food.

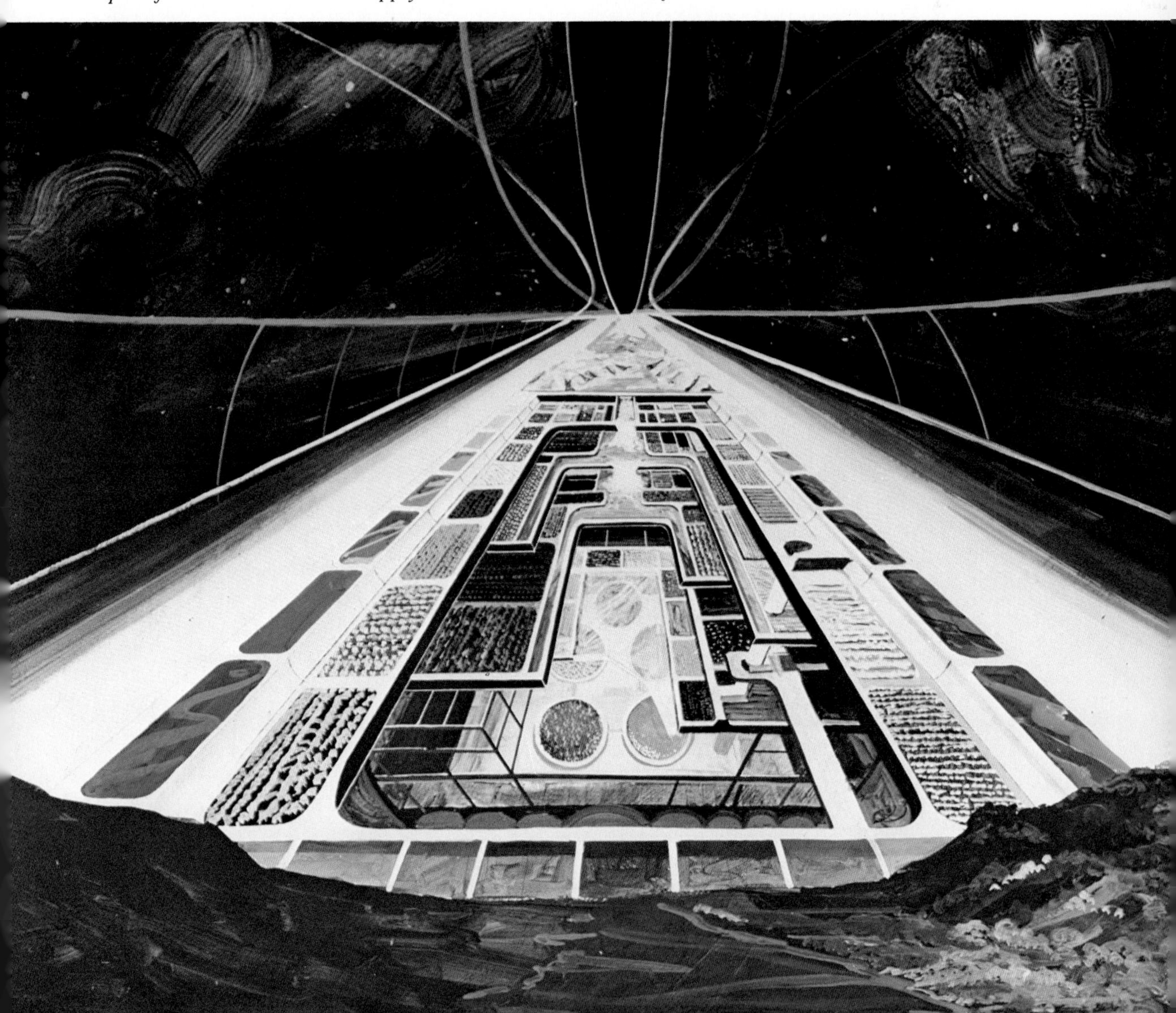

space compared to what it costs to launch them into space; dropping into the gravity well takes much less effort than climbing out of it. Lunar gravity-assists would help deorbit the food packages, making the impulse required infinitesimally low. And the technology for safely landing large quantities of food aboard lightweight vacuum-foamed reentry bodies is well understood. As we have seen, astronomers Michael Gaffey and Thomas McCord have explored similar techniques for recovering asteroidal metals on earth.

How much asteroidal material must be retrieved? And how much would a mission cost to build a growing area big enough to feed the world? In *The High Frontier*, Gerard O'Neill estimated that food grown in space on an area of about one-hundredth of a hectare would easily feed one person. Therefore, for a projected world population of six billion in the year 2000, the growing area requirement would be more than one-half million square kilometers. Assuming that the soil and structural mass of an agricultural pod are about ten centimeters thick (with most of this mass in the form of water harvested from the asteroids), we would need about 50 billion tons of mass for space farms to feed the world. A three kilometer diameter asteroid would be adequate. As pointed out in an earlier chapter, the solar system is peppered with thousands of asteroids that size or larger.

Estimates of the cost of retrieving this quantity of material are subject to a number of uncertainties, but its magnitude can be determined by examining engineering and economic studies of satellite power-station construction and asteroid retrieval during the early phases of a program of processing nonterrestrial materials. The cost per unit of weight required for the larger-scale space agriculture operation (involving *billions* rather than *millions* of tons) would be proportional to the size of the powerplant of the asteroid-retrieving rocket. Assuming a first generation solar-satellite powerplant, the final cost of retrieving the asteroid would be in the range of 100-500 billion dollars.

Costs will also be incurred in the processing of the materials needed for manufacturing the growing areas, the shipment of phosphorus from the earth, the construction of deorbiting rockets with metal-foam heat shields to deliver the food offshore from receiver countries, the towing of the food ashore, and local distribution.

A useful analog to the recovery of asteroidal material for growing food in space is Gaffey and McCord's scenario described in the last chapter, for recovering on earth eight billion tons of iron and nickel from a one kilometer diameter asteroid. Transport and processing costs were estimated to be about $140 billion per year for the recovery of about 300 million tons of iron and nickel. World food productivity is about 200 million tons per year, with a potential marketplace of several hundred billion dollars per year (food production in the United States is currently valued at $100 billion per year).

Gaffey and McCord have proposed that 100,000-ton payloads of iron and nickel ore could be safely landed aboard a "flattened shuttlecock" reentry body made of lightweight, floatable metal-foam material fabricated in space. A gentle, low-energy nudge of these payloads from a manufacturing orbit high above the earth along a trajectory past the moon and thence a reentry path to the earth would be all that is required to deorbit the ore. The same principle would apply to the delivery of some 800,000 tons of food harvested in space each day to offshore sites on the earth to match projected world food production shortly after the year 2000. Engineering studies have verified the feasibility of the steps required to recover, on earth, food and materials originally obtained from an asteroid.

The capital-investment-plus-interest of some hundreds of billions of dollars in the food-growing areas in space is comparable to the 700 billion dollar capital cost estimate by Revelle for irrigation development and agricultural modernization in the Third World. More importantly, the potential annual market would be between $200 billion and $500 billion per year, so the payoff would be rapid, even assuming a high rate of discount on the investment. In terms of potential markets and expanding the limits to growth on the earth, there are some striking similarities between this concept and O'Neill's concept of providing baseload electricity worldwide from satellite power stations constructed from nonterrestrial materials, as well as Gaffey and McCord's concept of recovering on the earth iron and nickel from asteroids. But qualitatively, the world food problem stands alone in its seriousness, for energy and metal resources have a greater potential for conservation, recycling, and renewal without causing human hardship, while food production must remain above a certain minimal level to prevent widespread famine. Meanwhile, the vicissitudes of climate, disease, and politics continue to prevent us from having a stable and lasting system of food production and distribution.

These considerations suggest that space agriculture, in its early stages, may provide a sort of "famine insurance" against a disaster on earth like the Bangladesh typhoon. In such a case, a more modest requirement for land area in space could be imposed (say, 100 times less) prior to scaling-up to earth-equivalent food production. A famine-insurance facility could come on line 15-30 years from now, by which time space enterprises would be growing rapidly. There already exist compelling arguments for rapid economic growth in space, and it would not be unreasonable to suggest that earth-equivalent food production in space might be achieved 20-40 years from now.

Nevertheless, it is unlikely that transporting food to the earth from space will provide the initial economic incentive for opening up the resources of space; it is more likely to be a natural byproduct of other programs early in the twenty-first century, unless a crisis precipitates a crash program. At least for the time being, space agriculture is not competitive with efforts to increase food production on earth, and that situation is unlikely to change over the next decade or two. Even then, it is likely that a blend of moral and economic factors will lead us to examine carefully all potential methods of increasing and stabilizing food production. The space solution is the most obvious ultimate method.

Although the aforementioned estimates are clearly preliminary, they seem to indicate that a potentially healthy economy, independent of the earth's biosphere, may permit the growing of a virtually unlimited supply of food for use on earth as early as 20 years from now. Such a possibility could relieve the immense pressures the human race now feels in managing its limited supply of nonrenewable energy, food, and other resources. The onset of space agriculture may occur just in time to avert massive starvation.

15. Colonies in Space

History will probably judge that the quest for energy, food, and materials was only a fleeting rationale for getting a purchase on the bounty from space. More significant will be the migration and eventual evolution of humans away from the mother planet. This door-opening will probably be no less significant than the first movement of our fish-ancestors onto land.

Although it is difficult to envision what the ultimate human habitats in space will look like, engineering and design constraints give us a pretty good idea of how the early ones might appear.

The standard design for an early space settlement is a rotating, 10,000-person, spherical habitat, one mile in circumference. As you walk from the equator toward the pole of the spin axis you get lighter and lighter. Imagine a one-tenth gravity swimming pool. Your feet are still pinned toward the outside but you weigh 10 to 20 pounds. That means you can jump over a house if you're agile. Imagine diving off a diving board, doing a few flips, and landing in the pool *above* you. Imagine what a one-tenth-gravity ballet would be like. We could all be Nadia Comaneci's at one-tenth gravity.

These things could happen only twenty years from now. Nadia Comaneci will be only 40 years old, and most of us will still be alive. Twenty years is plenty of time. Recall that President Kennedy set the lunar-landing goal in 1961, and just eight years later the landing was accomplished, far sooner than anybody believed possible.

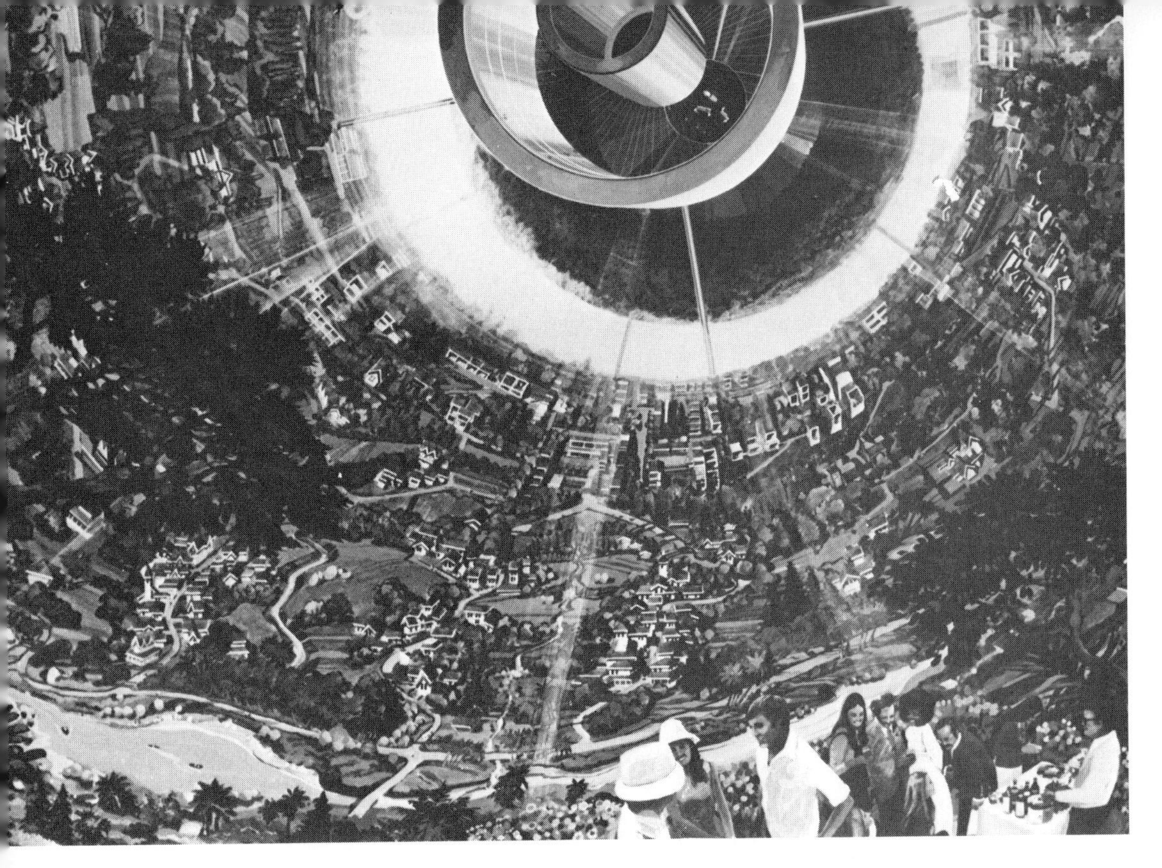

(Above) Following a tradition established at the NASA Ames Summer Studies, inhabitants in this space settlement are enjoying a wine tasting party. (Below) These space colonists prefer a tropical environment, as swimmers test their skill diving at one-tenth gravity (top).

(Right) Later space colonies like this one can house millions of people; in this view we are looking down the long axis of a cylinder several miles long containing alternate panels of land and glass.

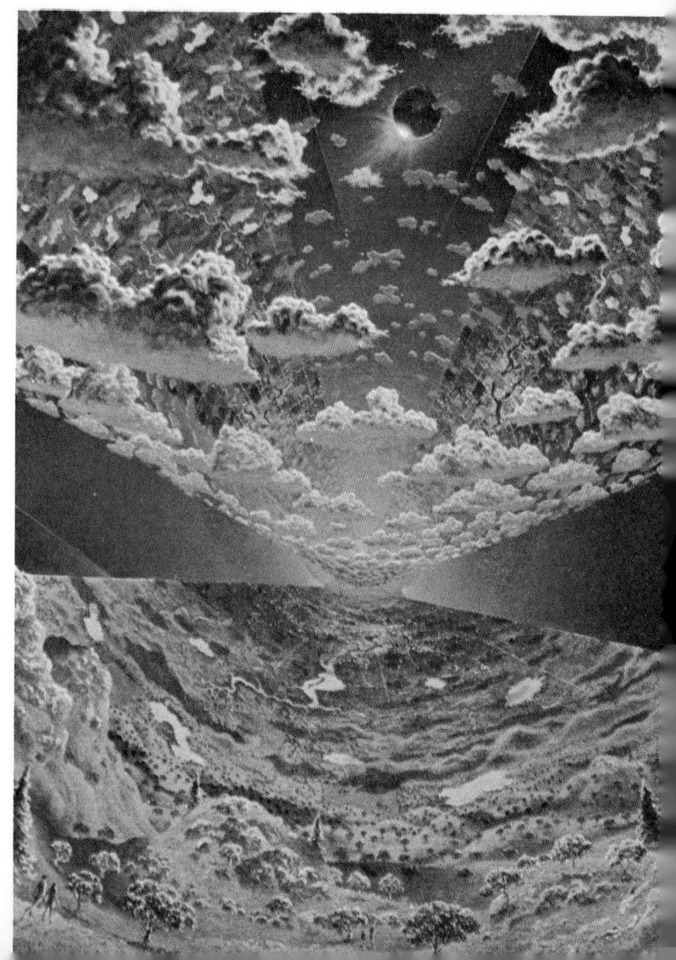

(Right) The outsiders of two huge co-rotating cylinder colonies are ringed by food-growing pods.

(Below) The inside of this space colony resembles the San Francisco Bay Area (the way it used to be).

Once millions of tons of asteroidal materials come in, very large settlements become possible. One design calls for cyclinders, a few miles in diameter and several miles long, rotating in opposite directions. They could house hundreds of thousands of people.

Civil engineering tells us that there is no reason why we can't build these structures as big as several miles across. If they are beyond that size, and we want to create one artificial gravity by spinning them, structural stresses caused by atmospheric pressure become too great. But if we can prove that people can live comfortably and in good health at a fraction of one gravity, then there is almost no limit to the size of a space habitat.

Solar energy is unlimited, and there is sufficient material in the asteroids to build enough habitats for ten thousand times the present population of the earth.

The inside of these large cyclindrical habitats could be earthlike. The spin-axis would point to the sun, and three panels of mirrors and glass would bring in sunlight and starlight. Moving the mirrors in and out would simulate the daily motions of the sun and stars. In one artist's view, a total eclipse of the sun by the moon is taking place, casting a shadow on the city in the background. Another artist fancies the San Francisco Bay area—the way it used to be. More exotic, unearthlike environments can also be created in space. Variable gravity and no gravity allow for an infinite array of three-dimensional designs and unprecedented recreational opportunities. Science fiction? No—space habitats are inevitable.

16. *Project Columbus 1992*

The surface of our earth has been thoroughly covered by man, and there is little left to explore. Until the beginning of the space age, mankind has had dwindling opportunities to satiate his exploratory instincts. As a result, the excitement of discovery has become available only to a small number of scientists with advanced degrees in a specialty, beyond the reach of the ordinary man and woman. The stifling of exploration places psychological stress on individuals and societies and increases the chance of hostility between countries in the face of dwindling resources and opportunities. Territoriality for a growing population on a finite earth becomes increasingly significant.

Over the last two decades, our eyes have opened to the planets in the solar system. Robots have been sent to the surface of Mars and have flown by Mercury, Venus, Mars, Jupiter, and Saturn. The missions have yielded spectacular results, increasing our understanding of the origin, evolution, and current environments of the planets. Dried-up riverbeds on Mars, huge volcanoes on Mars and Io, the deep, swirling atmospheres of Jupiter and Venus, the red, arid, rocky surface of Mars, and the esthetic beauty of the Voyager pictures of Jupiter, Saturn, and their satellites—all these provide tantalizing visual previews of what is to come: firsthand human exploration of the planets.

These other worlds are "crying out for exploration," as Carl Sagan pointed out, but during the 1980s NASA will virtually shut down its planetary program. The November 1980 Voyager encounter with Saturn has been called "The Last Picture Show" by Jet Propulsion Laboratory director Bruce Murray. The headlines and turmoil of our times distract us from the fact that the past two decades will probably be best remembered as the era when mankind first glimpsed the moon and planets close up.

The 1980s promise to be one of the most dismal, unimaginative periods in the human history of new physical discovery. Iran, Afghanistan, presidential elections, increasing oil prices, gas-station lines, Watergate, Vietnam, airline crashes, terrorism, nuclear accidents, protests, blackouts, blizzards, floods, droughts, hurricanes, tornados, earthquakes, bomb explosions, assassination attempts, murders, and suicides dominated the headlines, television news, and conversations of the 1970s. These repetitions of human frailty and natural disaster come and go like fashions, but they will be short paragraphs—or footnotes—in the history books of the future, even though they are all consuming issues now.

What will a twenty-first or twenty-second century historian make of the last third of this century? Which particular events are *unprecedented* and will change the course of human history? Reaching and utilizing the fertile stars would undoubtedly be an event of unprecedented magnitude. It would profoundly change the direction of history—and it could happen before the arrival of the new millennium.

What is sometimes difficult to realize is that we are now involved in events that are unprecedented, historically indelible, and could lead to human exploration of the solar system. At worst, if we were to abandon space exploration, which is a highly unlikely prospect, future history books are certain to be filled with pages of description and photographs of mankind's first journeys to other worlds during the past 15 years. Besides the transient thrills of the manned flights during the 1960s, the first robot-camera glimpses of those exotic, beautiful, and scientifically curious new surroundings have been the stuff of page 3 in the newspapers and 15-second spots buried in TV news programs.

History will not ignore these events. Man has landed on the moon and brought back soil from its surface. Robots have landed on Mars and Venus and have flown by Mercury, Venus, Mars, Jupiter, and Saturn, snapping pictures and measuring temperatures, compositions, geology, magnetic

(Above) Artist's concept of the Mariner 9 orbiting Mars with the Earth and Moon in the background. (Left) The U.S. Viking lander took this photograph of the rock-strewn, ruddy Martian landscape.

(Opposite page: top) The Pioneer Venus orbiter took this picture of the cloud-enshrouded sister planet of Earth. (Bottom) Jupiter's cloud patterns taken by Voyager resemble a modern painting.

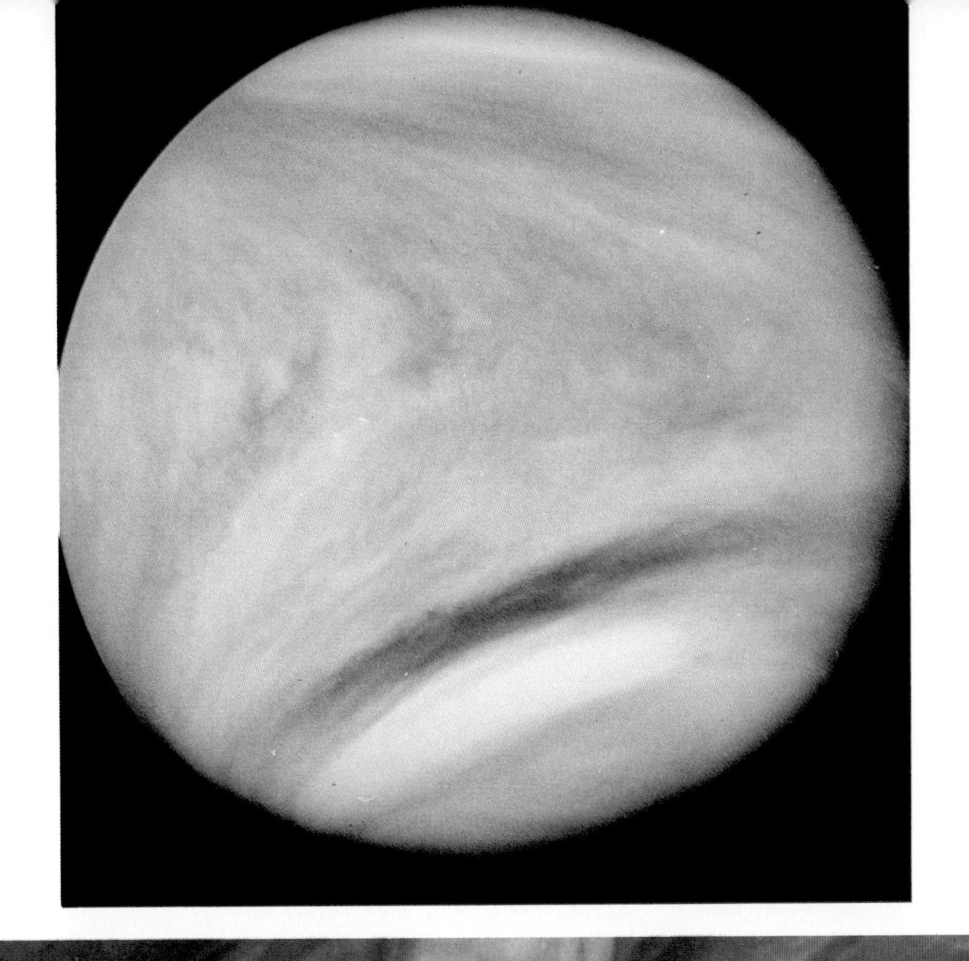

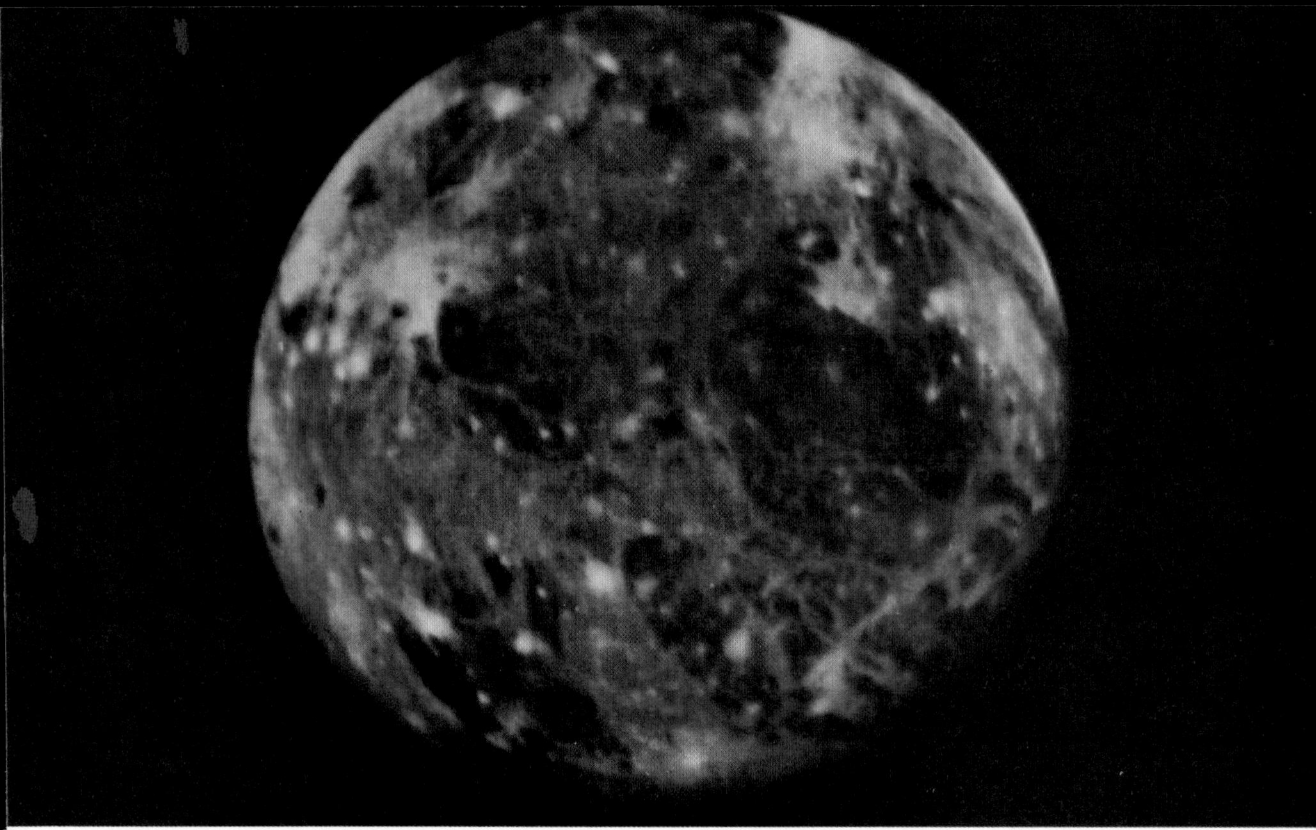

(Above) Craters surrounded by rays of ice excavated by meteroid impacts dominate the landscape of Jupiter's largest satellite, Ganymede. (Below) The fourth large satellite of Jupiter, Callisto, resembles our moon.

(Opposite page, top) Volcanoes and weird fields of sulfur dominate the landscape of Io, Jupiter's pizza moon. (Opposite page, bottom) Crevices and cracks traverse an otherwise smooth Europa, Jupiter's second big moon.

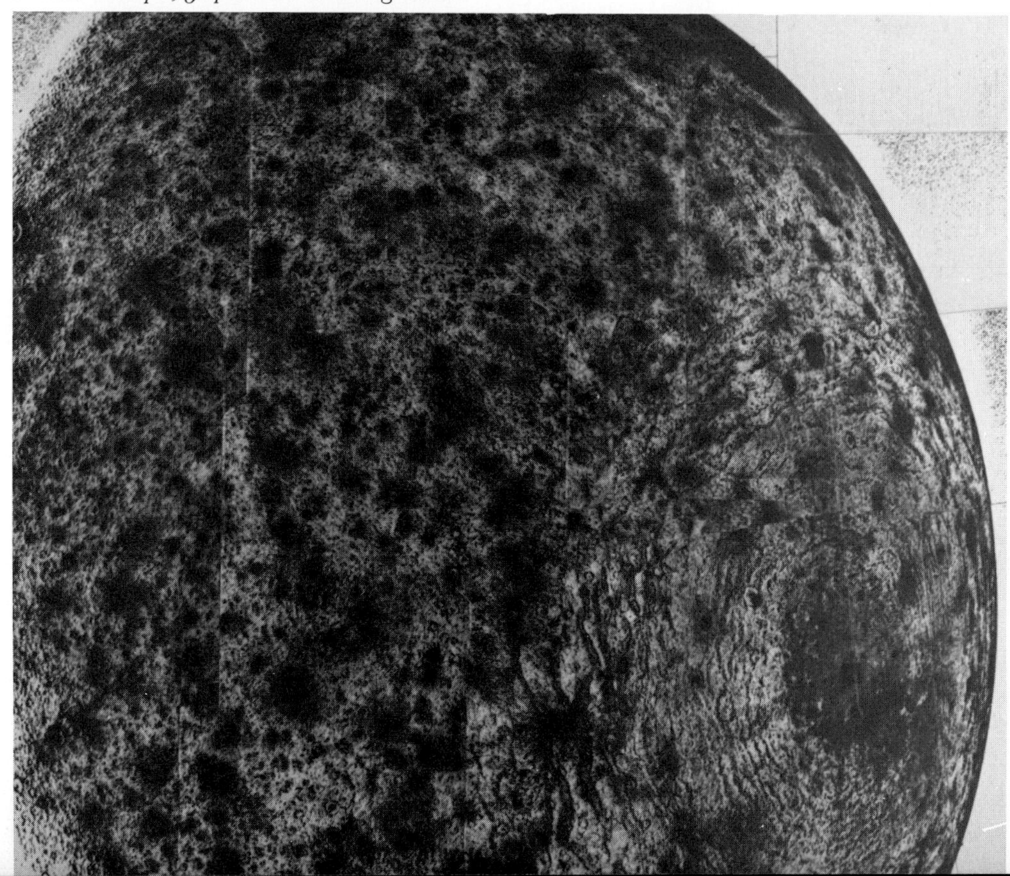

fields, atmospheric motions, and a host of other data. The results have been astounding. For the first time in history, our eyes are open to new worlds, each one a unique environment that developed independently over the eons.

The 1976 photographs of the surface of Mars taken by Viking cameras show a boulder-strewn desert resembling the outskirts of Tucson, Arizona. The Viking mission produced some interesting and seemingly contradictory data concerning the question of life on Mars. The two Viking landers sampled only a tiny fraction of the surface of the Red Planet, and we will probably be unable to resolve the life debate until the new era of exploration that is a natural outgrowth of space manufacturing of nonterrestrial materials.

Viking photographs taken from orbit around Mars show dried-up river beds, evidence that liquid water flowed some time in the history of the planet. (Life as we know it requires liquid water for survival.) The atmospheric pressure on Mars is now too low to support liquid water, yet it was once there. Perhaps these river beds will provide some fossil evidence of former life on Mars.

We cannot answer these questions—and many more—without an aggressive program of planetary exploration. As historically significant, scientifically exciting, and esthetically pleasing as the Mariner, Pioneer, Viking, Voyager, and Soviet planetary missions have been, there are many more new questions unanswered than old questions answered. We are compelled to find out more.

Following the euphoria of the Apollo program, it was fashionable to talk about an Apollolike 100 billion dollar manned Mars-landing, a one-shot affair, but the money and interest weren't there. More valuable, in my opinion, would be a thorough exploration of the solar system, based on a firsthand, multifaceted, and continuing scientific effort like Cousteau exploring undersea or Darwin's Beagle expedition to the far corners of the world. The most likely pathway to a thorough exploration of the planets and moons lies in opening the fertile stars. Ships of exploration fabricated from the moon or asteroids will make all this possible by the year 2000 and be far less expensive than a one-shot Earth-to-Mars Apollo analog.

For more than a century, the terrestrial globe has held diminishing opportunities for the explorer. The solar system offers extraordinary new opportunities and discoveries that may make the first forays to the Western World 500 years ago seem like drives down the street. What more appropriate way would there be to celebrate the quincentennial of the discovery of the New World than by embarking on an exploration of the new worlds of the solar system? Project Columbus 1992 would be the equivalent of thousands of Project Apollos, a sustained and systematic exploration of the solar system by man and machine made possible by latching onto the fertile stars.

What might the first Project Columbus explorers see on their journeys to the planets and moons? They could visit the gigantic Martian volcano Olympica Mons, three times the height of Mt. Everest. Mars also has its own Grand Canyon, but it is three times deeper than ours and it would span the continental United States. A planet of superlatives, this other world developed separately from the earth for the last four billion years and has evolved its own esthetic and scientific identity, which, after so many eons, is ripe for exploration.

University of Virginia physicist S. Fred Singer has suggested that Phobos could be a base of manned exploration of Mars. Its low gravity and abundant resources could provide a comfortable, radiation-proof haven for future astronauts. This idea dovetails very well with the economic rationales for developing the resources of space, and Phobos is likely to be inhabited by humans before too long.

The mysterious planet Venus, forever enshrouded with clouds, has a surface pressure of 100 atmospheres and a surface temperature hot enough to melt lead. Explorers will need an airtight, heatproof bathyscape to wander about the surface of Venus. Someday it will happen.

Hot, cratered, moonlike Mercury would be a strange place to visit. Because of its slow rotation and eccentric motion around the sun, there are spots where you would see the sun set in the west, rise again where it had set, and then, some hours later, set again.

There is probably no more colorful and exciting place to visit than Jupiter and its moons. It is a gigantic gas ball, almost as big as a star, and its cloudtops appear as turbulent belts, bands, and spots. Most prominent is the Great Red Spot, a permanent storm bigger than the earth. Voyager photographs revealed beautiful pastel swirls that would attract attention as a first-rate modern painting. Nature has done it again! (Part of our geocentric chauvinism is that the earth is the *only* beautiful planet, but we are beginning to find out otherwise.)

Jupiter's four largest moons (of at least 13) rate high as exotic places. In 1610, when the Italian astronomer Galileo looked through the newly invented telescope at Jupiter, he noticed four starlike specks orbiting the giant planet. These Galilean satellites helped establish that the earth was not at the center of the solar system. Although it was heresy at the time, Galileo supported the Copernican view that the planets orbited the sun and that some of the planets had their own moons, as if they were suns with their own solar systems.

The four Galilean moons (Io, Europa, Ganymede, and Callisto) are as different from one another as the planets are. They are each about the size of our moon, but that's where the similarity ends. Voyager pictures of the innermost one, Io, reveal a bizarre red, white, orange, and yellow volcano-pocked moon, somewhat resembling a pizza. Active volcanos abound on Io, one appearing as an umbrellalike plume rising 270 kilometers above its surface. Io, not the earth, is the most active known geological object in the solar system.

Europa, the next moon out, looks like a smooth, yellowish-white billiard ball, crisscrossed by cracks thousands of kilometers long. Again, we see evidence of a lot of internal activity, but the surface is probably composed largely of ice.

Jupiter's largest moon, Ganymede, also has some cracks, some mottled regions of various hues, and craters like those on our moon, Mars, and Mercury. The ejecta blankets from the bigger and younger craters are white, suggesting subsurface ice.

Callisto also shows signs of ice, but it has large, ancient impact-basins similar to our moon's and Mercury's, revealing a primitive, unaltered surface.

A trip to any of these moons would be an extraordinarily exciting event, full of scientific and esthetic potential. On Io, for example, one could observe an enormous red, white, and orange oblate Jupiter in a multicolored sky with the other moons moving to-and-fro through the heavens.

Pioneer—and now, Voyager—are beginning to unlock the scientific mysteries of Saturn and its rings and moons. They will probably prove to be as unique and exciting as Jupiter's system. Saturn's moon Titan is bigger than the planet Mercury. What lies beneath its thick atmosphere? And what further mysteries await us around Uranus, Neptune, and Pluto?

The U.S. and Soviet programs of gradual, systematic planetary exploration are likely to continue to provide tantalizing glimpses of other worlds in our solar system. But only a Project Columbus, based on dividends from the fertile stars, will enable us to build the truly large, comfortable ships that will carry us directly to these exciting destinations. We are on the threshold of a new era of human exploration, far beyond our dreams, and it is in our power to do so *as a byproduct* of the economic incentives that are driving us to the moon and asteroids.

Beyond the planets, large telescope arrays in space will be able to probe the mysteries of the universe with far greater sensitivity than is possible on earth. The riddle of extraterrestrial intelligence may be unlocked by the detection in space of radio signals. Enormous radio and optical telescopes will be easy to build in space as a byproduct of space manufacturing. The day may come, perhaps within a generation, when starships containing families will head for the greatest adventure of all: exploring the great unknown surrounding the 100 billion stars in the Milky Way galaxy.

PART V

Should We or Shouldn't We?

17. *Gumption Traps*

Whenever there are good reasons to do something, usually also there are reasons why we should not. In some cases, the negatives outweigh the positives for purely rational and humane reasons, for example, murder, theft, rape, embezzlement, and suicide. In our quest for the fertile stars, evaluating the negatives is more complicated, but most of them concern the resistance common to *any* radical new perspective that may change the course of human affairs.

Approaching this resistance, overcoming it, and surrendering to change requires a special effort—often in response to a desperate need or crisis—that often transcends common sense and logic. Robert M. Pirsig, in the book *Zen and the Art of Motorcycle Maintenance*, refers to this transcendence as *gumption*.

The Apollo program took gumption. The roots of the Apollo decision were firmly implanted in 1961, just after President Kennedy had made a serious tactical blunder in ordering Amerian troops to invade the Bay of Pigs in Cuba. The invasion failed and embarrassment ran deep in American souls. The Soviets had been the first in space, both with machines and with men. "Are we losing the Cold War?" was a frequently asked question in the United States at that time. Kennedy desparately groped for a dramatic show of American strength, and the lunar landing was his choice.

On May 25, 1961, President Kennedy said in a message on urgent national needs to Congress, "Now is the time to take longer strides, time for a greater new America, time for this nation to take a clearly leading role in space achievement, which in many ways may hold the key to our future on Earth. . . . I believe that this nation should commit itself to achieving the goal of landing a man on the moon and returning him safely to earth."

Neither Kennedy nor his top advisers knew exactly how the goal was to be achieved, although they had a good idea that it could be done, provided a big enough rocket were built and a large technological infrastructure were established. Skeptics reacted with disbelief, and controversy abounded.

No matter, the impossible had become possible: a collective human feat of awesome proportions, motivated by emotion. Was it worth doing? Of course it was, because it felt right at the time. Stopping the discussion of the pros and cons of the Apollo decision at that point may seem to some now to have been an intellectual cop-out, but I argue that it was not an intellectual decision subject to analysis. It was an emotional decision—and the right one—a *fait accompli* once Kennedy and Congress acted, disregarding the extent of technical challenge and the cost.

We now face a challenge analagous to Apollo. What is stopping us is feelings of timidity and retrenchment, of resistance, of experiencing what Pirsig calls a *gumption trap.* Setting aside narrow reasons for not making the giant leap into space (possible microwave pollution, vulnerability to military attack, etc.), it is interesting to investigate the broader gumption traps. These include bureaucratic stultification, industrial shortsightedness, lack of money, international treaties, overgrowth, the "small-is-beautiful" philosophy, and overcentralized control.

It may be all too obvious, but it is still worth stating that the key elements in space exploration are politics and emotion, not technological detail and scientific results. Politics motivate the technology to make it all happen. The dramatic and clear intention to land men on the moon was the crucial step, actually 99 percent of the battle. The technology to accomplish it was merely the mechanics—challenging, yes, but only the visible tip of an enormous iceberg built of the human qualities of hope, fear, courage, anger, passion, love, hate, and determination. Money was no object. The Apollo success was the result of a synergistic, warlike mobilization of collective human emotion. The severe crew-cuts, the bewildering acronyms, and the cool shop-talk between mission control and outer space projected an incongruous image, diametrically opposed to an underlying emotional excitement of historical proportions.

Ironically, just as the space agency's post-Apollo II euphoria began, public opinion sagged. Only weeks after July 20, 1969, the same public that had glued its eyes for hours to Walter Cronkite narrating the dramatic adventures of Armstrong, Aldrin, and Collins, now regarded any subsequent space activity with a blasé attitude. The public feeling at the time was that Apollo II was a hard act to follow, and we now have problems to deal with here on earth.

Do you remember Vice President Agnew's goal of a manned Mars-landing by the end of the century? It went over like a lead balloon. Do you remember the names of the crews of Apollos 12 through 17? Of Skylab? Of Salyut? Forced by public apathy to reduce its ambitious post-Apollo goals, NASA was left with the space shuttle, a budget-trimmed, marginally well-designed vestige of more grandiose visions. It is a legacy of Apollo, a historical accident; yet, ironically, it is an essential stepping-stone to the fertile stars. NASA has staked everything on the shuttle, while it is understandably concerned about its own survival. That concern invites a policy of fear-reaction, job-survival calcification and a refusal to accept new ideas (even though NASA is full of secret enthusiasts, anxious to embrace new opportunities).

Meanwhile, jurisdictional disputes between NASA and the Department of Energy have delayed consideration of the nonterrestrial sunsat option, and the bureaucracy slogs along in its multiple considerations and gumption traps. U.S. spending on space studies and experiments is equivalent to a few minutes worth of defense spending.

Despite all this waffling, there are encouraging signs. In a speech made by NASA Administrator Robert Frosch in the fall of 1979, he said, "It appears possible to start with the investment necessary to put 100 tons of

machinery on the moon and after 20 years of machine reproduction to have an energy plant and manufacturing capability equivalent to the ability to manufacture 200 billion pounds [100 million tons] of aluminum per year. . . . I believe that the technology is presently available [to build such self-replicating machines] and that the necessary development could be accomplished in a decade or so."

The private sector is waiting for the U.S. government or some other government to act. While the utilities, the aerospace industry, and large manufacturing companies anticipate trillions of dollars coming in during the first few decades of the twenty-first century, the relatively few billions needed for investment is still too much for any one company—which needs to justify its annual investments and profits to its shareholders—to risk so far in advance. Bruce Hannay, Vice President for Research at Bell Telephone Laboratories, put it this way: "I see the research scene on a shorter horizon. As a result, there is a decline in effort on major innovation that produces an entirely new class of technology or new class of service." This loss of the edge in technological leadership by the United States cannot help but further weaken our economy. While the oil dollars flow out at an increasing rate, the high-technology dollars flow in at a decreasing rate.

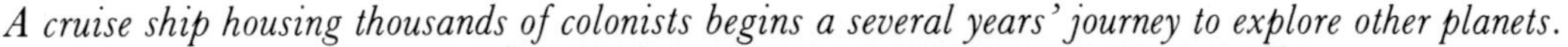

A cruise ship housing thousands of colonists begins a several years' journey to explore other planets.

If neither government nor industry in the United States has the gumption, perhaps some other government will prod the world into an Apollo-like Cold War mobilization. The Soviet space program is quietly moving toward the establishment of a manned space station. Where will they go after that? The Russians may be behind the United States in building a space shuttle, but they are ahead of us in designing human and agricultural habitats in space. They continue to launch humans into orbit; they hold the duration record in spaceflight; they have landed unmanned vehicles on Venus; and they have a vigorous military space program.

The oil-rich Arab nations, Japan, and Western Europe also have the awareness and the resources to begin to act. What seems to be missing is a triggering mechanism or crisis to fan the flames. Meanwhile, the world wallows in gumption traps. The less-developed countries would also like to have a piece of the action and do not want to be subjugated to even greater economic disadvantages. These nations support the Moon Treaty, which would prevent any private company from expropriating nonterrestrial materials and would make these resources available to all countries. To some individuals, this treaty represents another disincentive to begin moving.

Other gumption traps occur on a more philosophical level. For instance, there is a debate dividing the environmental community concerning growth versus no-growth, small-is-beautiful, and decentralization. The growth/no growth debate can be summarized as follows:

Growth advocates believe that a rapid transition to no-growth is not possible with the existence of poorer nations seeking to raise their standard of living to that of industrialized nations. The transition, they argue, is contrary to the tenets of existing political systems, and it would appear to be impossible, on purely practical grounds, for at least several decades. They suggest, too, that a steady-state world would impose severe restrictions on individual freedom and opportunity.

The steady-state advocates, on the other hand, believe that growth must stop to ensure human survival. Taking the limits-to-growth models at face value, even allowing for some argument about precise dates in the scenario, the no-growth argument holds that political systems and lifestyles must adapt to these limits or suffer the consequences, like any biological system that overbreeds and runs out of prey. There seems to be little dispute that this reasoning is valid when applied on a global scale.

Others have attempted to reconcile these views by acknowledging the seriousness of the problem and proposing long-term earth-based solutions. There is, for example, Emile Benoit's dynamic-equilibrium economy. which advocates a shift in growth from the depletion of nonrenewable resources to the rapid development of renewable resources, the elimination of waste, the reduction and eventual stabilization of population growth, and the reliance on science and technology to provide innovations for an improved quality of life. MIT economist Jay Forrester has argued that more attention needs to be paid to social rather then physical limits. E. F. Shumacher has proposed that technologies and human activities need to be developed on a small scale, with accountability by individuals and small groups rather than the large, amorphous organizations that currently control our destiny.

Whether these measures are politically possible or would be adequate to turn the tide is open to debate. There is some skepticism that these proposed solutions could be realistically implemented before widespread disaster strikes. We have seen that the space solution to the limits to growth transcends the dilemma on a time scale of centuries. The terrestrial limits to growth of food, energy, material resources, pollution, and population could be relieved in a matter of years after the first retrieval of materials from the shallow gravity wells of the moon or the asteroids.

I believe that the most serious challenge in using the resources of space is the danger that uncontrolled exploitation and economic gain could further upset the social order and create new difficulties on a massive scale. An exploitation of space that merely provides mankind with relief from the limits of the biosphere could result in a one-way growth into the solar system; it would merely buy time before we came up against new limits. It's easy to envision profit-hungry exploitation creating energy overdemand or a relaxation in dealing with problems here on earth.

In other words, the solution to one critical human problem, i.e., shifting consumption from nonrenewable resources on earth to renewable and nonterrestrial resources in order to raise the standard of living of poor countries, might help create a new problem: large-scale exploitation and a relaxation of standards.

These considerations should provide ample warning that an international political framework for controlling the exploitation of space must be developed. About 50 million tons of material—an excavation on the moon one kilometer square and 15 meters deep, or a 300 meter diameter asteroid—is all that is needed to construct enough power satellites to supply the world with all its energy in the year 2000. Later retrievals of much larger quantities of asteroidal materials could provide the earth with abundant food and metal resources. Eventually it may be desirable to establish large colonies for developing exploratory opportunities and alternative lifestyles.

In any case, given the possibility of a desperate situation here on earth in the relatively near future, I believe that the best hope for mankind lies in a vigorous effort to establish something like Benoit's dynamic-equilibrium economy *and* space manufacturing and asteroid retrieval. To hesitate on the basis of vague fears of runaway growth and environmental deterioration, in my opinion, would be unwise and contrary to human nature.

18. Economic and Ecologic Incentives

We are in a recession, and the stage is set for further decline. The evidence for this is overwhelming: a balance of payments imbalance ($50 billion of oil imported by the United States each year, which is likely to increase for at least the next several years); decreasing productivity in the United States; increasing scarcities in natural resources; inflation; and overcapitalization in outmoded technologies, e.g., the automobile industry. The doomsday literature is full of dire prophecies about what might happen as a result: increasing international tensions; energy and food famines; brownouts and blackouts; severe cold; nuclear war; irreversible pollution; etc.

The evidence is strong that we will continue in this decline through the 1980s, and perhaps into the 1990s, and recover afterward. Forrester, for example, argued that new initiatives must await a decline in the use of today's overbuilt capital equipment. He foresees a deep economic drop from which we will begin to recover during the 1990s, consistent with long-range (Kondratieff) cycles in the economy that take place every 50 years or so.

Deforestation, pollution, carbon dioxide buildup, radioactive releases, strip-mining, and the danger of nuclear war are among the many sources of concern environmentalists have expressed about our future. The overwhelming consensus is that the planet is seriously threatened by environmental neglect.

There is good news, however. The space solution to the limits to growth transcends the dilemma. We have seen that the terrestrial limits to growth could be relieved in a matter of years after the first retrieval of materials from the asteroids. Large-scale industry could eventually move into space, removing pollution pressures from the earth. Population stresses could be relieved by the option to live in space. The space solution handles all five major parameters studied in limits-to-growth models: energy, food, materials, population, and pollution. In principle, the earth could turn into a bountiful garden during a prolonged era of abundance and prosperity.

By 1990, space industrialization will have progressed far enough for the space solution to be clear and tangible. The investment will have been small compared to those for conventional energy production and weapons. Economic studies of satellite power stations indicate that if they were constructed from lunar and asteroidal materials, the earth's energy problem would be solved. Even if this energy source were not to work out, there exist other alternatives that will inevitably point to the economic use of the resources of near-earth space: nickel and platinum from the asteroids, and titanium from the moon; the construction of large-scale military satellites; and the public demand for shuttle passenger service.

As we pull out of the recession of the 1980s, these activities in space will be a natural part of the recovery, just as our mobilization during World War II followed the depression of the 1930s. A series of negative events during the 1980s is likely to trigger the space solution. History has shown us that events will probably then take place more rapidly than we anticipate. Some people argue that because the resources of space are so vast, there will be no more Kondratieff dips, just a rising curve for 100 or more years.

Freeman Dyson sees space as the site for a reconciliation between the conflicting values of the ecologist's small-is-beautiful philosophy and the capitalist's and government's big-is-necessary position. In his book *Disturbing the Universe,* Dyson posited two types of technologies that will help solve our problems: green (small, environmentally compatible, low-technology) and gray (big, centrally-controlled, high-technology). Rooftop solar collectors, solar ponds, and fuels made from plants are examples of green technologies. Nuclear power, the Apollo program, satellite solar power, and large space colonies are examples of gray technologies. Dyson believes that both approaches will continue to serve mankind, and that it is oversimplistic to judge green as good and gray as bad.

Following the ideas put forth by Konstantin Tsiolkovsky, Dandridge Cole, and Gerard O'Neill, Dyson envisions groups of adventuresome families homesteading the asteroids. Unlike the gray, centrally-controlled, monolithic space colonies (NASA-style), he believes that small-scale asteroid settlements will become the new green frontier. He argues that the investment in homesteading an asteroid would be less, as a percentage of gross national product, than that of the Pilgrims' Mayflower expedition or the Mormons' first settlement of Utah, due to the removal of bureaucratic overhead and the pioneer's willingness to take greater personal risks.

19. *The Quality of Life*

The bad news of the previous chapters provides overwhelming evidence that we are entering a frightening and depressing period of history. There is little available evidence that this will change in the near future.

Societies behave like individuals externalized, so it is interesting to see how individuals are handling the increased stresses of our time. The answer appears to be not well. Families are breaking up at an increasing rate, and the children of broken marriages—which is turning out to be *most* marriages—will be the leaders of the early twenty-first century. The lack of proper nurturing does not bode well for these people. The high-anxiety, myopic world we live in is not conducive to the development of strength of character, imagination, self-confidence, and enjoyment. The result may be resignation to mere survival or even mental illness.

The good news is that all of this will turn around rapidly at the dawn of the postterrestrial age. "The Earth is the cradle of Man, but we cannot stay in the cradle forever" were the words of Konstantin Tsiolkovsky. Our generation includes the adolescents of the postterrestrial age. The space solution will evolve from a project geared to solving mankind's most serious problems into a positive new spiritual and philosophical thrust into a territory that promises to be immensely stimulating, much like the Renaissance.

Some of our foremost theologians, philosophers, and psychologists recognize the spiritual significance of what is to come. Timothy Leary,

Barbara Marx Hubbard, and others have explained that the forthcoming human migration into space will result in fundamental evolutionary changes as profound as those experienced by the first animals that moved from sea to land. The potential for those parts of the human brain, they explain, that are now elusive or inaccessible to us could be released, engendering great new insights. Significant changes could take place in a span of just a few generations of space-bred humans. According to some studies, life extension or even physical immortality are probable consequences.

Short of these speculations (which I believe are bound to happen), some astronauts have experienced spiritual insights or transcendental feelings while in space. Even down-to-earth scholars have remarked on the significance of the dawning of a new age in space. Eugene Kennedy, professor of psychology at Loyola University, used the astronaut's view of earthrise above the lunar horizon as a symbol of the "dawning of a new spiritual awareness."

In an interview with Joseph Campbell, considered to be the foremost American scholar in the field of mythology, Campbell said, "The mystical theme of the space age is this: the world, as we know it, is coming to an end. The world as the center of the universe, the world divided from the heavens, the world bound by horizons in which love is reserved for members of the in-group: that is the world that is passing away. Apocalypse does not point to a fiery Armageddon but to the fact that our ignorance and our complacency are coming to an end. . . . It's fashionable now to demand some economic pay-off from space, some reward to prove it was all worthwhile. Those who say this resemble the apelike creatures in *2001*. They are fighting for food among themselves, while one separates himself from them and moves to the slab, motivated by awe. That is the point they are missing. He is the one who evolves into a human being; he is the one who understands the future.

"There have been budget cutbacks in the space program. We shrug it off. But that is where we live. It is not out there. . . . The fear of the unknown, this free-fall into the future, can be detected all around us. But we live in the stars and we are finally moved by awe to our greatest adventures. The Kingdom of God is within us. And Easter and Passover remind us that we have to let go in order to enter it."

British essayist Henry Fairlie, a critic of virtually every twentieth-century technological adventure in the western world, supports this view. "It is my claim," he writes, "that science has now returned to us the possibility of cosmology, that once again we can freely find ourselves, and our ethics and politics, under the 'dome' of a universe that we can see and understand for ourselves. . . . The universe is infinite and therefore the

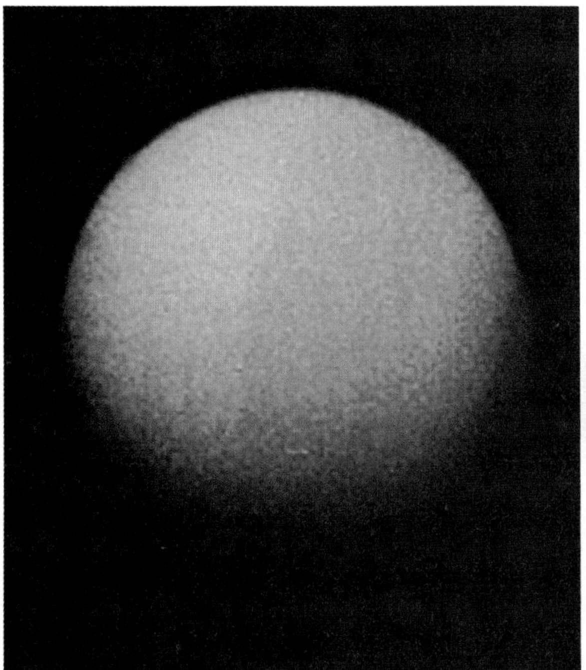

"It is time for us to realize that we are too great a nation to limit ourselves to small dreams"
—President Ronald Reagan, Inaugural Address, January 20, 1981.

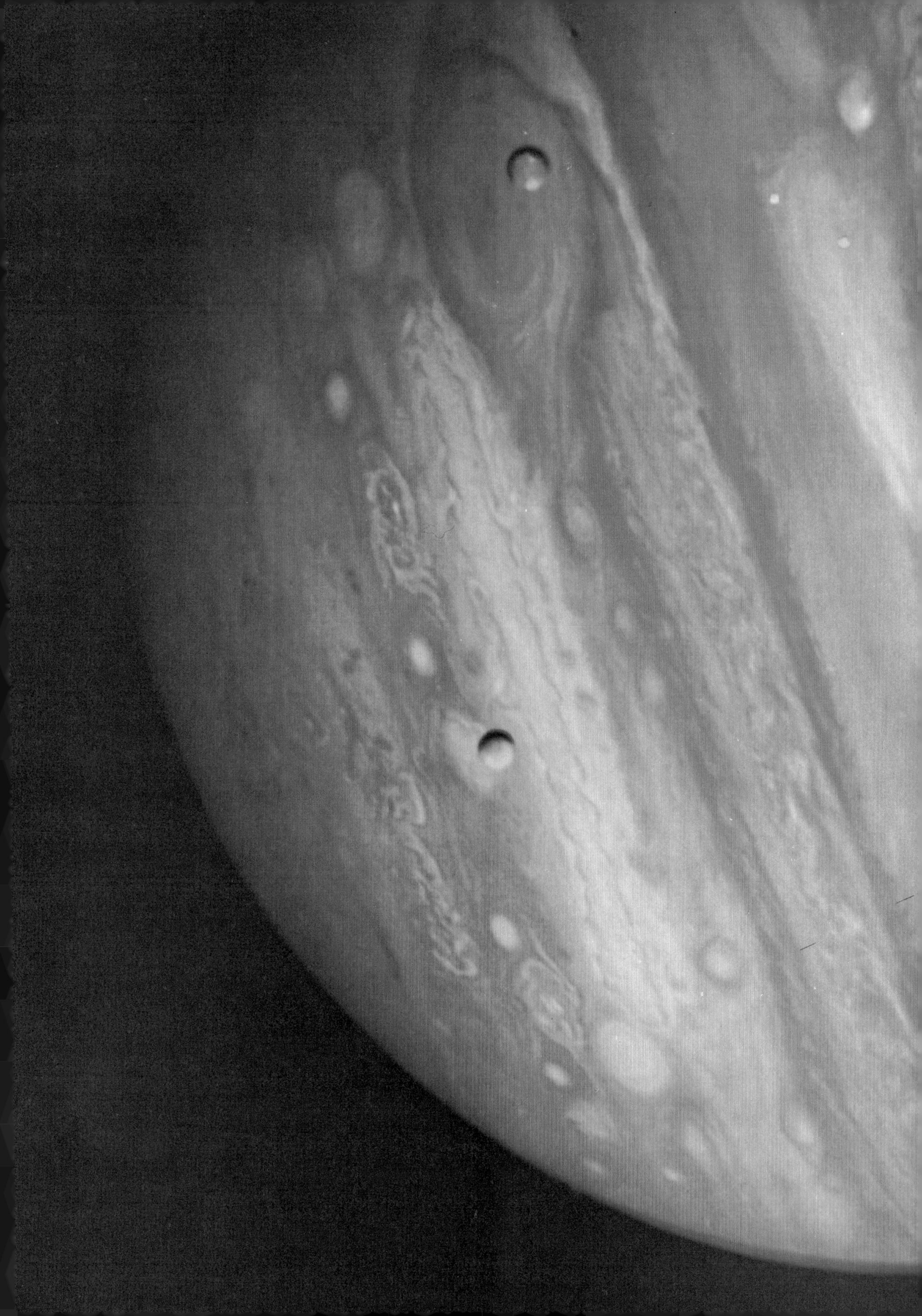

future is open. It will not be long before the space shuttles are up there and from then it will not be long, as a serious scientific writer has persuasively envisioned, before most of mankind is also out there, returning to the earth as we now go to Athens, in order to see 'where it all started.' "

"The conservative mood of our century," Fairlie continues, "and more recently of this decade, has been pitted against our science as well as against nature. The contemporary conservative is anxious and nervous about science. He does not know where it will lead, on the one hand; on the other hand, he feels that it will lead to a lack of reverence. But it is precisely this need for reverence, even for some form of transcendence, that our science is more and more willing to encompass. . . . Our world and the future are not closed, but are opening to us as never before. This is the new vitality for which we, our philosophy and our societies and politics, have waited so long."

Psychiatrist Jerome Frank sees our nuclear age as a source of deep fear. In order to alleviate that fear, he argues, there would need to be "a world order with effective institutions for prevention of all wars. A psychological prerequisite to the achievement of this distant, perhaps unattainable goal would be a sense of world community among all peoples, transcending national loyalties." He cited "the more rapid exploration of outer space" as an example of an activity which could lead to the international unity so urgently needed.

20. *The PostTerrestrial Age*

Mankind is beginning to take the first steps out of the cradle of earth, and they are painful. Most of us are unaware of what is happening in space, what *will* happen in space, and the profound influence it will have on all of us. Space shock will intensify the pain and anxiety for many of us already experiencing negative feelings during the 1980s. Some individuals, companies, and countries will use the space solution to further their own powerful ends. This calls for a new international responsibility and order. There will be hardships and risks for the first pioneers in the space frontier.

We can turn the shock around and prepare to share in the exciting new adventure ahead. There is plenty to go around. Unity of mankind with himself and with the universe are the next logical steps toward his fulfillment and the removal of his deepest anxieties. These spiritual, philosophical, psychological, and social dimensions transcend the nuts-and-bolts economic logic and Cold War political fervor that ushered us into the postterrestrial age. Deep down, we feel a oneness with the cosmos, and we are compelled powerfully and inevitably toward the heavens. We are destined to be up there, although perhaps at first not for the "right" reasons. It is in our blood.

Participation in the postterrestrial age will be like a beautiful calm after a fierce storm. The growing pains of the 1980s, and perhaps also the 1990s, and the crises that we shall encounter may be more severe than any ever experienced in human history. But just as the pain will be intense, so, too, will be the sense of relief. Humanity is on the verge of the ultimate rollercoaster ride, but we will recover, breathlessly, and enter the realm of the stars.

In the spring of 1957, while a senior in high school in Belmont, Massachusetts, I was asked by my history teacher to write a paper on an important contemporary topic. I chose the subject of space satellites, and the teacher wrote on my paper, "well researched, but not relevant." Sputnik was launched the next fall, and the world has not been the same since. Perhaps my history teacher would have preferred a treatise on the modern implications of the Missouri Compromise, for he was (understandably) one of the majority of pre-Sputnik geocentric thinkers. If, during the summer of 1957, one had asked people in the street whether they believed we would go to the moon in twelve years, the chances are they would have dismissed the interviewer as being crazy.

My main point is that major technological advances take the world by storm; they are unanticipated even by the most respected futurologists. The public perception of space exploration in the ten years since Neil Armstrong took his first small steps on the moon seems to have reached record levels of ennui. This year, a high-school student's desire to write about space colonies or the search for extraterrestrial intelligence might evoke the same resistance I experienced.

Today, most of what we hear about space is the fall of Skylab, cost overruns, scheduling delays of the space shuttle, and Soviet charged particle-beam ABM satellites. But an increasing number of scientists, engineers, and laypeople are becoming aware that waiting in the wings are concepts of space exploration and utilization that stagger the imagination and that could be implemented on a large scale, using technology that is now available within manageable budgets. These events could occur within the next 20 years, an interval less than the span of the space age, bringing us to the year 2000, when unimaginable space odysseys could take place.

The question is not *if* we do it, but *when* do we do it? Such a prediction is subject to speculation, just as it was during the pre-Sputnik era. When the door opens—as a result of emotion, not bureaucratic logic—I am certain that events will proceed at a staggering pace. The fulfillment of nearly every conceivable wild idea is ours for the asking during most of our lifetimes.

The triggering mechanism will be a blend of a political or economic crisis and strong, visionary political leadership. Perhaps a nuclear power plant will melt down. Maybe a long, crippling oil embargo will take place. Maybe the Soviet Union will orbit a permanent space station or cause a nuclear war. Perhaps life will be found in an unlikely place, such as the surface of one of Jupiter's moons. Whatever the event may be, it is now conceived of as unlikely to happen—as unlikely as the results: industries and colonies in space, first-hand exploration of the planets, enormous radio and optical telescope arrays in high orbits, asteroid mining, starships, etc.

The need to plan for 10, 20, and 30 years from now is essential. The stakes are too high to be tentative in exploring the space-manufacturing and asteroid options. Regardless of the outcome of debates about growth, energy policy, and political philosophy, the evidence is clear that research and development of space-based manufacturing from nonterrestrial materials should begin immediately. The inevitable alternative is growing international inequities in energy and food supply, more oil embargos, expensive energy, depletion of nonrenewable resources, proliferation of dangerous nuclear technology, massive hunger and starvation, and tighter controls over human freedom.

Starting the logical progression of steps required to use the vast resources and energy available to us in space demands a unity, discipline, and perspective now lacking because of a collective tentativeness about trying something so totally new. The challenge is more one of communication than of technical feasibility or economics. History makes it clear that this project will go ahead sooner or later. The arguments are compelling that this is the time to begin to fulfill some of mankind's highest hopes.

"What we need is a Copernican revolution, and we don't have it," said Dennis Meadows, author of *The Limits to Growth.*

We *do* have it.

INDEX

Numbers in italics indicate illustrations